Springer Theses

Recognizing Outstanding Ph.D. Research

Aims and Scope

The series "Springer Theses" brings together a selection of the very best Ph.D. theses from around the world and across the physical sciences. Nominated and endorsed by two recognized specialists, each published volume has been selected for its scientific excellence and the high impact of its contents for the pertinent field of research. For greater accessibility to non-specialists, the published versions include an extended introduction, as well as a foreword by the student's supervisor explaining the special relevance of the work for the field. As a whole, the series will provide a valuable resource both for newcomers to the research fields described, and for other scientists seeking detailed background information on special questions. Finally, it provides an accredited documentation of the valuable contributions made by today's younger generation of scientists.

Theses are accepted into the series by invited nomination only and must fulfill all of the following criteria

- They must be written in good English.
- The topic should fall within the confines of Chemistry, Physics, Earth Sciences, Engineering and related interdisciplinary fields such as Materials, Nanoscience, Chemical Engineering, Complex Systems and Biophysics.
- The work reported in the thesis must represent a significant scientific advance.
- If the thesis includes previously published material, permission to reproduce this must be gained from the respective copyright holder.
- They must have been examined and passed during the 12 months prior to nomination.
- Each thesis should include a foreword by the supervisor outlining the significance of its content.
- The theses should have a clearly defined structure including an introduction accessible to scientists not expert in that particular field.

More information about this series at http://www.springer.com/series/8790

Satomi Shiraishi

Investigation of Staged Laser-Plasma Acceleration

Doctoral Thesis accepted by
The University of Chicago, USA

Author
Dr. Satomi Shiraishi
The Enrico Fermi Institute
University of Chicago
Chicago, IL
USA

Supervisor
Prof. Young-Kee Kim
The Enrico Fermi Institute
University of Chicago
Chicago, IL
USA

ISSN 2190-5053 ISSN 2190-5061 (electronic)
ISBN 978-3-319-35928-1 ISBN 978-3-319-08569-2 (eBook)
DOI 10.1007/978-3-319-08569-2

Springer Cham Heidelberg New York Dordrecht London

Softcover reprint of the hardcover 1st edition 2015

Printed on acid-free paper

Springer is part of Springer Science+Business Media (www.springer.com)

Parts of this thesis have been published in the following journal articles:

A. J. Gonsalves, K. Nakamura, C. Lin, J. Osterhoff, S. Shiraishi, C. B. Schroeder, C. G. R. Geddes, Cs. Tóth, E. Esarey, and W. P. Leemans, *Plasma channel diagnostic based on laser centroid oscillations,* Phys. Plasmas **17**, 056706 (2010).

A. J. Gonsalves, K. Nakamura, C. Lin, D. Panasenko, S. Shiraishi, T. Sokollik, C. Benedetti, C. B. Schroeder, C. G. R. Geddes, J. van Tilborg, J. Osterhoff, E. Esarey, Cs. Tóth, and W. P. Leemans, *Tunable laser plasma accelerator based on longitudinal density tailoring,* Nature Phys. **7**, 862 (2011).

C. Lin, J. van Tilborg, K. Nakamura, A. J. Gonsalves, N. H. Matlis, T. Sokollik, S. Shiraishi, J. Osterhoff, C. Benedetti, C. B. Schroeder, Cs. Tóth, E. Esarey, and W. P. Leemans, *Long-Range Persistence of Femtosecond Modulations on Laser-Plasma-Accelerated Electron Beams,* Phys. Rev. Lett. **108**, 094801 (2012).

G. R. Plateau, C. G. R. Geddes, D. B. Thorn, M. Chen, C. Benedetti, E. Esarey, A. J. Gonsalves, N. H. Matlis, K. Nakamura, C. B. Schroeder, S. Shiraishi, T. Sokollik, J. van Tilborg, Cs. Tóth, S. Trotsenko, T. S. Kim, M. Battaglia, Th. Stoehlker, and W. P. Leemans, *Low-Emittance Electron Bunches from a Laser-Plasma Accelerator Measured using Single-Shot X-Ray Spectroscopy,* Phys. Rev. Lett. **109**, 064802 (2012).

S. Shiraishi, C. Benedetti, A. J. Gonsalves, K. Nakamura, B. H. Shaw, T. Sokollik, J. van Tilborg, C. G. R. Geddes, C. B. Schroeder, Cs. Tóth, E. Esarey, and W. P. Leemans, *Laser red shifting based characterization of wakefield excitation in a laser-plasma accelerator,* Phys. Plasmas **20**, 063103 (2013).

To my families and friends

Supervisor's Foreword

Particle accelerators have been key drivers for a broad spectrum of fundamental discoveries and transformational scientific advances since the early twentieth century. Each generation of particle accelerators builds on the previous one, raising the potential for discovery and pushing the level of technology ever higher. The science and technology of particle accelerators developed for particle and nuclear physics research have transformational applications for (i) other areas of basic science such as atomic physics, bio-physics, and material science (using synchrotron light sources and spallation neutron sources) and (ii) many areas of benefit to the nation's well-being including medical isotope production, cancer treatments, biomedicine and drug development, national security (scanning of shipping containers), food sterilization, power transmission, and nuclear waste transmutation. It has evolved to become an important engine driving world economy. Over 30,000 accelerators are in operation worldwide, primarily for industrial and medical purposes.

Experimental investigation of laser-driven plasma accelerators (LPAs) is a topic pushing the limits of physics and technology. An LPA uses plasma as a medium to transfer a laser energy into the kinetic energy of charged particles. This novel concept offers the potential to reduce the size of accelerators by a factor of a thousand. However, LPAs require a number of advancements before they can be used reliably. One critical milestone is particle acceleration using multiple LPA units sequenced one after another (staged acceleration). Studies of the physics critical to realizing staged acceleration are the topics of Satomi Shiraishi's Ph.D. thesis.

Experiments were conducted using the 40 TW laser system at the LOASIS facility at Lawrence Berkeley National Laboratory. Design, installation, and commissioning of the setup were an important part of her thesis work. From these experiments, she produced high-quality electron beams, measured slice energy spread and emittance, and characterized plasma wave amplitudes. These topics form the basis of staged acceleration by improving beam quality and by understanding accelerating fields and electron beam dynamics. The anticipated demonstration of staged acceleration is an exciting beginning for LPAs to further develop toward the next generation of compact accelerators.

The thesis work of Dr. Satomi Shiraishi is to perform fundamental accelerator science, providing the foundation in knowledge and workforce upon which major advances in accelerator-driven technologies will be based. It conducted research with capabilities that would change existing paradigms, advance accelerator science at a fundamental level, and develop transformational applications in the crosscutting academic discipline. It could enable discoveries that lead to novel, compact, powerful, and/or cost-effective accelerators.

Chicago, March 2014 Prof. Young-Kee Kim

Foreword

Laser-plasma accelerators (LPAs) are an exciting area of current research owing to their ability to sustain ultrahigh accelerating gradients, of the order of 100 GV/m, some three orders of magnitude beyond conventional radio-frequency linear accelerators. Experiments on LPAs at Lawrence Berkeley National Laboratory (LBNL) have been successful in generating electron beams in the GeV range using a 40 fs laser pulse with a few joules of energy interacting with a plasma channel of a few centimeters in length. With further development, it is hoped that LPAs can provide the next generation of particle accelerators with applications ranging from high energy physics to advanced light sources, to compact accelerators for basic science and medicine.

A key component of LPA technology is the plasma channel, which serves to guide the laser pulse as well as to sustain the large amplitude plasma wave responsible for accelerating the particles. At LBNL, we are using a plasma channel technology based on capillary discharges, in which the plasma is generated within a narrow cylindrical channel machined through a block of sapphire. Because sapphire surrounds the plasma channel, it is difficult to use conventional plasma techniques to measure the plasma profile and to diagnose the laser–plasma interaction. Measurement of the plasma profile and diagnosis of the laser–plasma interaction is essential to both the understanding and optimization of LPA.

As part of her Ph.D. dissertation, Satomi has developed a novel technique to diagnose the laser–plasma interaction within the plasma channel. In brief, this is accomplished by measuring the properties of the laser pulse exiting the plasma channel, just after interaction with the plasma, and comparing this with the properties of the laser pulse entering the plasma channel, before interaction with the plasma. By measuring how the plasma interaction has modified the laser pulse, important properties of the laser–plasma interaction can be ascertained. For example, by measuring the amount that the laser pulse wavelength has increased due to the laser–plasma interaction, Satomi was able to determined how much laser energy was deposited within the plasma as well as how large a plasma wave was generated within the plasma channel. Knowledge of these quantities—laser energy deposition and plasma wave excitation—is essential to understand and develop LPAs. The techniques developed by Satomi, the data she obtained, and her analysis of the LPA physics within the plasma channel are truly impressive.

In addition to her outstanding scientific achievements, Satomi has been a pleasure to mentor and work with these past several years. I am very proud of her accomplishments. She has authored several key papers, given excellent presentations, was awarded the best student presentation at the 2010 Advanced Accelerator Concepts Workshop, and the 2013 Yodh Prize from the University of Chicago. I have no doubt in my mind that Satomi will continue to develop as an exceptional researcher and that she will have a very successful career, no matter what area of science she chooses to pursue. I wish her the best in her future endeavors.

Berkeley, May 2014 Dr. Wim P. Leemans

Acknowledgments

This Ph.D. was part of a longer journey which would not have been possible without the support of many teachers, colleagues, family, and friends over the years.

I was extremely fortunate to have Prof. Young-Kee Kim as my advisor and mentor at the University of Chicago. Young-Kee, you inspired me and have been a wonderful role model as a person and as a physicist. Thank you for making time for me and helping me explore my choices. Our Saturday morning meetings sharing a fresh pot of coffee were valuable; your expectations were high, but you always provided me personal support and helped me grow professionally and personally. I will always be grateful for your guidance.

I feel very lucky for the opportunity to have worked in the world's leading laser facility with Dr. Wim Leemans as my advisor. Wim, thank you for having confidence in me to perform experiments, letting me explore my ability, and nudging me to believe in what I can do. I learned a lot through your leadership for work and for fun, and I enjoyed being a part of LOASIS. Thomas, thanks for sharing your experience with me on staging. We worked hard together and I appreciated your teaching. Tony, Kei, Jeroen, Brian, Sven, Cameron, Nicholas, Danny, and Joost, thanks for your guidance, support, and encouragement throughout my study. Our work was often intense, but you knew how to make it fun. I learned about physics, about people, about teamwork, and a lot more. Carlo, Carl, Eric, and Jean-Luc, thank you for your help in my theoretical understanding; knowing the physics always made it more interesting to work on our projects. With the construction of the staging setup, technical and engineering support was indispensable. Don, Dave, Mark, Ken, Nathan, Greg, Ohmar, Art, Rob, and Dennis, thanks for your professional support on the experiment. Your support always extended beyond the task at hand, and I learned so much working with you. I also want to thank everyone else in LOASIS for making my graduate experience rich and enjoyable.

My friends kept my mind and body happy and healthy. Roger, Ioana, and Andrea, thank you for everything you have done especially when I had my back problem; I knew everything would be fine when I had you around. I am grateful that you took each step of my Ph.D. process beside me, being happy for good news and providing me with another way to see the situation when the news was difficult. I also want to thank Sachiyo, Sonja, Miguel, Ajilai, Krystyna, Sonia, and Susan for their friendship, and Julia and the Prestige family for helping me balance work with my love for horseback riding.

The past 15 years of my life in the U.S. could not be spoken of without my American family. Mary and Alex Platz, and Claire, thank you so much for taking me in as a part of your family. From the time I could not speak English and could only respond to your questions with a smile, you have supported me and helped me grow.

I feel deep appreciation to my grandparents, Shigemi and Kiyomi Kataoka, and Fusako Shiraishi, for sharing their life experiences with me. Their strengths and stories inspired and motivated me to pursue physics and to keep thinking about what is important in my life.

The support from my family has been incredible. To my parents, Tadashi and Mariko, I want to say how grateful I am for your trusting me and being so supportive of my choices. Being away from you and moving to the U.S. was hard at times, but you always made sure I was cared for. I also know that this journey was not easy for you either. Thank you for always thinking what is best for me. My brother and sister have been guardians and friends. Itaru, thank you for taking care of me. You bought me my first laptop, took me on trips, and made sure that my life was balanced. Chihiro, your international life has enriched my life and I am grateful for all the love you and Asheesh have given me (not to mention all the DVDs!). Everyone in my family was indispensable. Without their guidance and support, I could not have achieved this Ph.D.

Lastly, I am grateful for the many people who have given me support and guidance in my life. I wish space would allow me to include all of your names here.

Contents

Chapter 1
General Introduction

1.1 Conventional Versus Plasma-Based Acceleration

Particle accelerators use electromagnetic fields to increase the energy of charged particle beams. From early X-ray tubes to modern synchrotrons, cyclotrons, and linear accelerators, the development of accelerators have involved many branches of science such as electromagnetism, solid-state physics, atomic physics, and plasma physics [1]. In recent years, applications of accelerators include material surface treatment, medical diagnostics, cancer therapy, food sterilization, as well as scientific research including biological, chemical, material science, particle and nuclear physics; they have been indispensable for scientific and technological advancements [2, 3]. Biological and chemical applications rely on generating GeV or multi-GeV electron bunches and converting the energy to X-ray or photon beam production. The Advanced Light Source at Lawrence Berkeley National Laboratory and the Advanced Photon Source at Argonne National Laboratory are examples of such accelerators. For particle physics at the energy frontier, ever higher energies are needed to explore new regimes. The Large Hadron Collider (LHC) built by the European Organization for Nuclear Research (CERN) was an accelerator commissoned in 2008 for particle physics research, designed to accelerate protons to 7 TeV in the accelerating ring $\sim$27 km in circumference. The proposed International Linear Collider (ILC) is designed to accelerate electrons and positrons up to 500 GeV. The proposed design length of the accelerator is $\sim$31 km, increasing the energy of particles by $\sim$35 MeV every meter [4]. While conventional accelerators can produce high energy particles, the increasing size and cost have made the realization of these high energy accelerators challenging.

Laser-plasma accelerators (LPAs) have the potential to be the next generation compact accelerators. LPAs have demonstrated high accelerating gradients on the order of tens of GV/m, two to three orders of magnitude greater than conventional accelerators. Conventional accelerators rely on technology based on radio-frequency (RF) electric fields sustained in a metallic cavity as illustrated in Fig. 1.1a. With this method, large accelerating structures are inevitable since the maximum electric

S. Shiraishi, *Investigation of Staged Laser-Plasma Acceleration*,
Springer Theses, DOI: 10.1007/978-3-319-08569-2_1

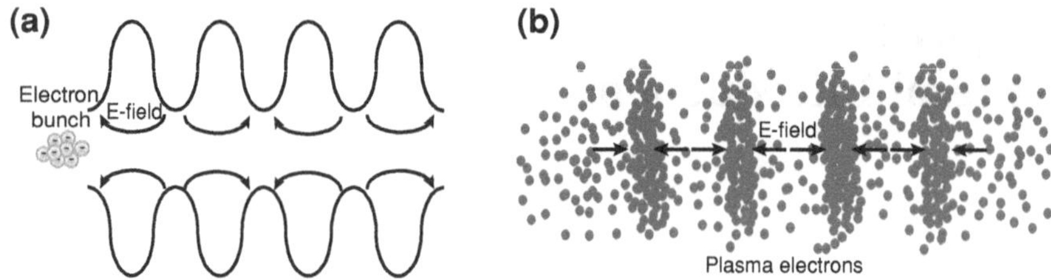

Fig. 1.1 Schematics of accelerating structures. **a** Conventional accelerator RF cavities where accelerating fields are sustained by charged metallic cavities. **b** Plasma-based accelerator where accelerating fields are sustained by local electron density modulations. With the plasma-based accelerator, the plasma is already ionized and can sustain a higher electric field than the metallic cavities

field sustainable in the cavity is less than 50 MV/m, due in part to the field-induced breakdown of the cavity walls. On the other hand, breakdown issues are not of concern in plasma-based accelerators because plasma is an ionized gas. The accelerating electric fields are sustained by a charge density modulation of plasma electrons as shown in Fig. 1.1b, and the accelerating fields can be on the order of 10–100 GV/m [5]. With these high fields, plasma-based accelerators can, in theory, achieve the desired particle energy in a much shorter distance than a conventional accelerator. This high accelerating field demonstrates the potential of the plasma-based accelerators to be the next generation of compact accelerators.

For most accelerating beam properties, the performance of conventional accelerators is still superior to that of current plasma-based accelerator technologies. Advantages of conventional accelerators includes a high repetition rate of 5–1000 Hz for linear accelerators and 10–500 MHz for circular accelerators. These high repetition rates allow for frequent particle collisions in the case of particle physics colliders and increased efficiency of radiation generation for light sources. Furthermore, the control of electron bunch properties such as charge, energy spread, and beam pointing stability are superior for conventional accelerators at this time.

However, plasma-based accelerators offer exciting possibilities for a new realm of applications through different beam properties and acceleration mechanisms. Plasma-based accelerators have demonstrated ultra-short electron bunch durations on the order of 10 fs [6–9]. These ultra-short electron bunches are thought to be necessary for the next generation of Free Electron Lasers producing hard X-rays [8, 9]. Furthermore in plasma-based accelerators, the electric fields are sustained in a plasma that is replenished after each acceleration. Even if the accelerating field is distorted with electron bunch propagation, a fresh supply of gas will form a new accelerator structure for the next shot. High accelerating gradient and ultra-short electron bunch duration, along with strong effort placed on improving the electron beam qualities, offer exciting possibilities for future plasma-based accelerators.

1.2 Laser-Plasma Acceleration

In plasma-based accelerators, the electric fields required to accelerate charged particles are sustained by electron density modulations in the plasma (ionized gas). Comprehensive overviews of plasma-based accelerators can be found in [5, 10]. Plasma waves in plasma-based accelerators can be excited by intense laser pulse(s) or by energetic particle beams. The scheme discussed in this thesis is referred to as laser-plasma acceleration (LPA) in which plasma waves are excited by the radiation pressure of a short, intense laser pulse ($>10^{18}$ W/cm^2). When an intense laser pulse propagates through a plasma, the radiation pressure of the laser pulse pushes away the free plasma electrons. These electrons are then pulled back by the ions which are considered to be stationary due to their heavier mass, initiating a charge density oscillation referred to as a plasma wave or a plasma wakefield [11, 12]. This oscillation has a characteristic frequency known as the plasma frequency, $\omega_\mathrm{p} = \sqrt{4\pi e^2 n_0/m_\mathrm{e}}$, where e and m_e are the charge and mass of an electron. The amplitude of the accelerating field is then estimated by $E_0 = c\, m_\mathrm{e}\, \omega_\mathrm{p}/e$ where c is the speed of light. For a typical plasma density of $n_0 \sim 10^{18}\,\mathrm{cm}^{-3}$, the maximum accelerating field is $E_0 \sim 96$ GV/m, three orders of magnitude greater than conventional accelerators [5]. This concept was first proposed by Tajima and Dawson in 1979 [13]. However, the technology for realizing the necessary laser intensities was not available until Strickland and Mourou demonstrated a terawatt laser system employing a chirped-pulse amplification (CPA) technique in 1985 [14]. In the CPA technique, a laser pulse is temporarily stretched to a lower intensity prior to amplification to avoid laser-induced damage in the gain medium. After the amplification, the laser pulse is temporarily compressed, producing an ultraintense laser pulse. This CPA technique made the experimental investigation of LPAs feasible. Since then, great progress has been made on LPAs both theoretically and experimentally. The concept of CPA and a more complete mathematical description of wake excitation are discussed in Chap. 2.

Objectives for the state-of-the-art LPA experiments include production of multi-GeV electron bunches, control of plasma waves, and optimization of electron bunch properties. The demonstration of 100 MeV level electron bunch production in 2004 by three groups [the Rutherford Appleton Laboratory (RAL) in the United Kingdom, Lawrence Berkeley National Laboratory (LBNL) in the United States, and the Laboratoire d'Optique Appliquée (LOA) in France] was followed by acceleration up to 1 GeV by LBNL in 2006 [6–8, 15, 16]. Now in 2013, pettawatt laser facilities such as the Texas Petawatt (University of Texas) and BELLA (LBNL) have achieved 2 and 4.25 GeV electron bunch production, respectively [17, 18]. Electron acceleration to even higher energies is expected in the near future. The control of electron bunch properties such as energy, energy spread and charge have been improved by tailoring plasma shapes and densities, by employing different gas species and by colliding multiple laser pulses [19–25].

However, even if the electron beam properties such as charge and energy spread are controlled, there is a limitation to the optimal energy gain of a single module LPA. One fundamental issue of an LPA is that the driving laser loses energy to excite plasma waves and eventually, the laser energy will be too low to continue the wake excitation. LPAs could overcome this depletion by using staged acceleration, a sequence of modules where each module is driven by a fresh laser pulse [26]. Other challenges include laser diffraction (a focused laser diffracts and reduces intensity as it propagates) and electron dephasing (accelerating electrons outrunning the plasma wave) [5]. For applications such as high energy accelerators, the LPA design will rely on controlling the laser diffraction and electron dephasing, as well as sequencing, or staging, multiple acceleration modules, each driven by its own laser to supply fresh drive laser pulses [27, 28].

This thesis focuses on experiments and theory relevant to staging laser-plasma accelerator modules. For all experiments discussed in this thesis, laser diffraction is mitigated by plasma channel waveguides. Extending the acceleration length by mitigating the laser diffraction in a plasma channel resulted in electron acceleration up to 1 GeV in 3 cm [15]. The control of electron dephasing is an active area of research and techniques have been proposed for the mitigation [29]. If a laser is guided, and electron dephasing controlled, energy gain is limited by depletion. To overcome laser energy depletion, staged acceleration is necessary. Experiments performed so far utilize a single laser that drives the wakefield for both injection and acceleration. With staged acceleration, injection and acceleration can be separated. Benefit of the staged acceleration is not only the supply of fresh laser pulses to LPAs but also the independent control of electron injection and acceleration. The on-going staged acceleration experiment at the LOASIS Program at LBNL will demonstrate driving two LPA modules with two independent laser pulses and the coupling of electron beams between them. A staged LPA requires understanding and control of electron injection, laser coupling into the beamline, and post-acceleration of the electron beams. This thesis focuses on the current progress of the staged acceleration experiment at the LOASIS Program.

1.3 About the Thesis

The work discussed in this thesis was performed at the LOASIS facility at LBNL. While I was a graduate student in the group of Prof. Young-Kee Kim at the University of Chicago, I spent four years at the LOASIS facility under the supervision of the Program Head, Dr. Wim P. Leemans.

This thesis will present the current status of the staged LPA experiment as well as challenges. Figure 1.2 illustrates the organization of the thesis. During the period of my study, the staging experiment setup was designed, built and commissioned. Throughout the thesis, laser transport lines followed by an e-beam line for the staging experiment is referred to as the staging beamline, and electron beams are referred to as e-beams. This thesis will present the initial results of experiments that will be the basis for staged acceleration. Electron injection in the 1st module (Chap. 4), laser pulse coupling with a plasma mirror (Chap. 5) and wake excitation in the 2nd module (Chap. 6) were independently studied. This thesis will also present the requirements and challenges of electron beam transport between stages and post-acceleration, based on parameters observed in the experiments.

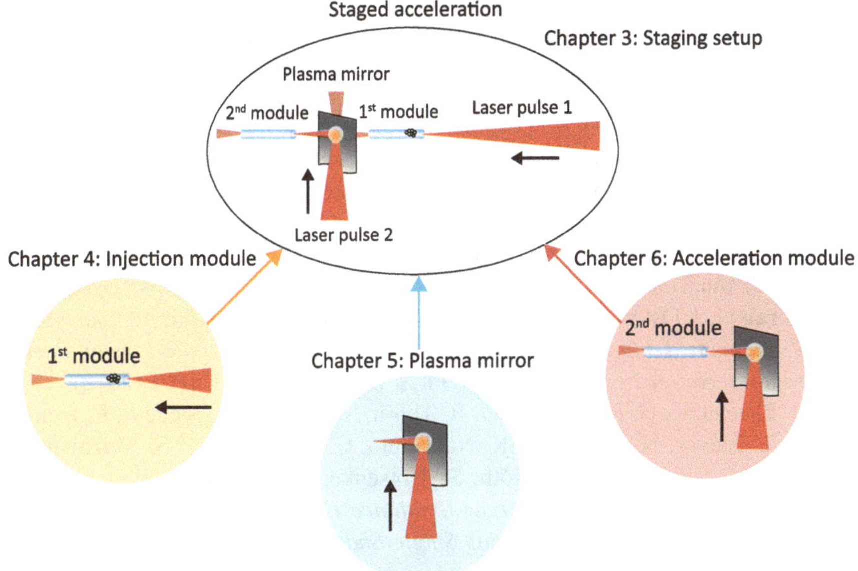

Fig. 1.2 Organization of the thesis. Chapter 3 will present staging experimental setup. Chapter 4 will discuss results from e-beam production study. Chapter 5 will discuss development and characterization of a VHS tape based plasma mirror. Chapter 6 will discuss a study of large wakefield excitation below electron trapping threshold

Chapter 2 will present the basic physics of an LPA. The concept of chirped pulse amplification and the architecture of the 40 TW laser system used in these thesis experiments are presented. The laser system is referred to as TREX. After the production of ultraintense laser pulses is discussed, the concepts of laser diffraction and guiding will be introduced. These laser propagation physics are crucial to understanding the laser evolution in the plasma channel. Then, the physics of plasma wave excitation, acceleration and dephasing of electron bunches, and several methods of electron bunch production will be explained. A discussion of the energy transfer from the drive laser to the accelerating electrons via plasma medium concludes this chapter.

Chapter 3 will be the introduction to the staging experiment at the LOASIS facility. The experimental goal and the design will be presented first. Then, the experimental configuration including the beamline layout and diagnostics implemented for the laser and electron beams will be described. Studies of the critical components for the staging experiment will be presented in the following three chapters: electron beam production, plasma mirror characterization, and wake excitation.

Chapter 4 will present experimental results on electron bunch production in LPAs which will be the basis of physics in the 1st module. The experiments were performed on the staging and the undulator beamlines which use the TREX laser system but are different in laser energies at the targets. The goal of this chapter is to demonstrate the various injection schemes introduced in Chap. 2 and to compare the electron beam qualities produced by different laser energies accessible with the TREX laser system. Requirements on the electron beam properties for successful transport between the stages will also be discussed. The discussion includes experimental results published in:

- A. J. Gonsalves, K. Nakamura, C. Lin, D. Panasenko, **S. Shiraishi**, T. Sokollik, C. Benedetti, C. B. Schroeder, C. G. R. Geddes, J. van Tilborg, J. Osterhoff, E. Esarey, Cs. Tóth, and W. P. Leemans, *Tunable laser plasma accelerator based on longitudinal density tailoring,* Nature Phys. **7**, 862 (2011).
- C. Lin, J. van Tilborg, K. Nakamura, A. J. Gonsalves, N. H. Matlis, T. Sokollik, **S. Shiraishi**, J. Osterhoff, C. Benedetti, C. B. Schroeder, Cs. Tóth, E. Esarey, and W. P. Leemans, *Long-Range Persistence of Femtosecond Modulations on Laser-Plasma-Accelerated Electron Beams,* Phys. Rev. Lett. **108**, 094801 (2012).
- G. R. Plateau, C. G. R. Geddes, D. B. Thorn, M. Chen, C. Benedetti, E. Esarey, A. J. Gonsalves, N. H. Matlis, K. Nakamura, C. B. Schroeder, **S. Shiraishi**, T. Sokollik, J. van Tilborg, Cs. Tóth, S. Trotsenko, T. S. Kim, M. Battaglia, Th. Stoehlker, and W. P. Leemans, *Low-Emittance Electron Bunches from a Laser-Plasma Accelerator Measured using Single-Shot X-Ray Spectroscopy,* Phys. Rev. Lett. **109**, 064802 (2012).

Chapter 5 will focus on the plasma mirror used to couple intense laser pulses onto the electron beam axis near the laser focus. This method will minimize the coupling distance between the stages to retain the overall high accelerating gradients in a staged LPA. The basic theory will be introduced to illustrate the physics. Then, the development and characterization of the tape-drive based plasma mirror will be presented.

Chapter 6 will present experiments relevant to efficient and controlled wake excitation in the 2nd module. All experiments presented in this chapter were performed below the threshold of electron injection and probe the physics important to acceleration modules. Characterization of a drive pulse, characterization of a plasma channel, and a diagnostic of excited wakefield will be presented. A laser pulse splitting process for the staging experiment will also be discussed along with an observed issue and means of mitigation for the issue. Initial experimental results and analysis on wake excitation in the 2nd module will be presented at the end. This chapter will include results published in:

- A. J. Gonsalves, K. Nakamura, C. Lin, J. Osterhoff, **S. Shiraishi**, C. B. Schroeder, C. G. R. Geddes, Cs. Tóth, E. Esarey, and W. P. Leemans, *Plasma channel diagnostic based on laser centroid oscillations,* Phys. Plasmas **17**, 056706 (2010).
- **S. Shiraishi**, C. Benedetti, A. J. Gonsalves, K. Nakamura, B. H. Shaw, T. Sokollik, J. van Tilborg, C. G. R. Geddes, C. B. Schroeder, Cs. Tóth, E. Esarey, and W. P. Leemans, *Laser red shifting based characterization of wakefield excitation in a laser-plasma accelerator,* Phys. Plasmas **20**, 063103 (2013).

Chapter 7 will present a summary of the work presented in this thesis. The future prospect of the staging experiment based on the initial experimental results will be discussed along with ideas on possible future improvements.

Chapter 2
Laser-Plasma Accelerators

2.1 Introduction

In this chapter, the basic physics of laser-plasma accelerators (LPAs) is introduced. In the LPA, electron properties such as energy, energy distribution, bunch duration, charge and divergence depend on the interplay between the drive laser and the plasma. The first part of this chapter focuses on the basic theoretical and experimental frameworks of the drive laser. The latter part of this chapter addresses mechanisms of plasma wave excitation, electron acceleration, and electron injection in an LPA.

Section 2.2 discusses the concept of chirped pulse amplification (CPA) laser systems often used in LPA experiments. Then, the architecture of the TREX laser system, a CPA based laser, is presented. Section 2.3 introduces the theory of laser propagation in vacuum which addresses laser diffraction. Laser diffraction, if not compensated, limits the acceleration in an LPA. Then, the theory of laser guiding in a plasma waveguide to mitigate diffraction is discussed.

Section 2.4 addresses the mechanism of plasma wave excitation and its properties such as plasma wavelength, amplitude, and phase velocity. It is illustrated that the properties of the plasma wave are mostly determined by laser intensity and plasma density. Electron acceleration by the plasma waves is discussed along with the concept of dephasing, which is another important aspect of the physics that can limit the acceleration length in an LPA. Then, methods of e-beam production will be discussed.

Section 2.5 describes the laser-plasma interaction in the context of energy transfer from the laser to electrons via the plasma medium. Scaling laws of electron energy gain in a single LPA module are discussed, which are limited by laser energy depletion if laser diffraction and electron dephasing are controlled. This laser energy depletion, also referred to as pump depletion, would not be a problem if the laser energy can be replenished. A staged LPA design has been proposed as a way to supply fresh laser pulses into the acceleration chain. Consequently, the staging of LPA modules will accelerate electrons to higher energy. Section 2.6 summarize the concepts introduced in this chapter.

S. Shiraishi, *Investigation of Staged Laser-Plasma Acceleration*,
Springer Theses, DOI: 10.1007/978-3-319-08569-2_2

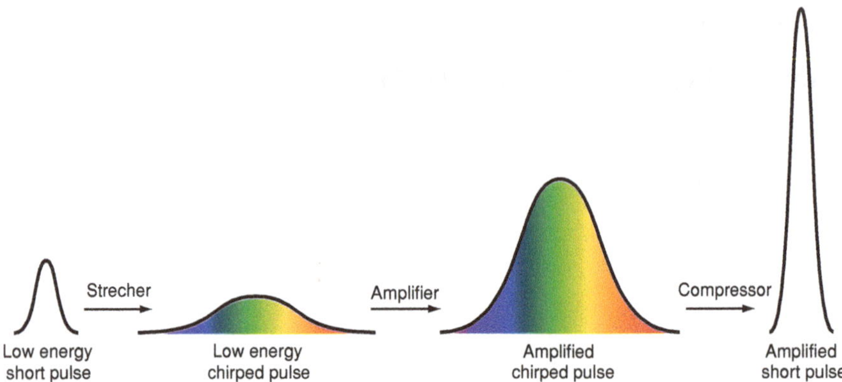

Fig. 2.1 A chirped pulse amplification system. A short, Fourier-limited laser pulse undergoes streching, amplification, and compression. The damage in the gain medium is avoided by temporarily stretching the laser pulse and reducing the intensity. The result of the CPA system is a high energy short laser pulse. When this pulse is spatially focused, an intensity of 10^{18} to 10^{20} W/cm^2 is achieved

2.2 Generation of Intense Laser Pulses

2.2.1 Chirped Pulse Amplification

Chirped pulse amplification (CPA) was the breakthrough technology that expanded the realm of experimental investigation of LPAs. Excitation of plasma waves rely on intense drive laser pulses above $>10^{18}$ W/cm^2 and duration $\tau \leq 100$ fs. A schematic of a typical CPA laser system is shown in Fig. 2.1 [14]. Initially, the relative phase of each frequency component is zero and produces a short pulse (5–10 fs). Prior to amplification, the pulse is temporarily stretched by the dispersion in the stretcher where the different frequencies take different path lengths. The resulting pulse is a long pulse (a few hundred pico-seconds) with linearly varying instantaneous frequency with time (color gradient in the figure). This linearly varying frequency is referred to as the chirp and the frequency is described by $\omega(t) = \omega_0 + \alpha\, t$ where α is the chirp rate. When $\alpha > 0$ (or $\alpha < 0$), the pulsed is said to be positively (or negatively) chirped. The stretched pulse has a low intensity compared to the damage threshold of the gain medium and can be amplified. The lasing medium used in most short pulse lasers for LPA research is typically sapphire crystals doped with titanium ions (Ti:Al_2O_3), pumped with green lasers of $\lambda \sim 532$ nm to set up the population inversion in the medium. When the seed laser (the stretched laser pulse) propagates through the lasing medium, the spontaneous emission of the excited atoms amplifies the pulse. Then, the amplified long pulse propagates through a compressor which reverses the dispersion in the stretcher. As a result, a short (a few tens of femto-seconds), high intensity pulse is produced. Peak power ranging from tera-watts to recently peta-watts can be generated enabling LPA experiments [30]. When the laser pulse is focused spatially by a focusing optic to 5–50 μm, intensity near the focus is on the order of 10^{18} to 10^{20} W/cm^2. These intense laser pulses are necessary in LPA experiments to excite plasma waves.

The compressor in the CPA system is often used to adjust pulse duration in LPA experiments. Changing the duration is a way to change laser intensity for a given energy. The shortest duration is determined by the spectral bandwidth of the laser. Since the concept of laser pulse duration and spectral width will be used in the analysis of some of the results presented in this thesis, the mathematical framework is presented below. The temporal profile of the complex laser field can be described with $U(t) = |A(t)| \exp(i\omega_0 t + i\phi(t))$ where $|A(t)|$ is the amplitude, ω_0 is the angular frequency and $\phi(t)$ is the phase. The instantaneous frequency of the laser pulse is ,

$$\omega(t) = \omega_0 + \frac{\mathrm{d}\phi}{\mathrm{d}t}. \tag{2.1}$$

This is related to the frequency domain via the Fourier transform with $\omega = 2\pi\nu$,

$$V(\nu) = \int U(t) \exp(-i2\pi\nu t)\mathrm{d}t \tag{2.2}$$

$$= |V(\nu)| \exp(i\psi(\nu)). \tag{2.3}$$

The optical intensity is $I(t) = |U(t)|^2$ and spectral intensity is $S(\nu) = |V(\nu)|^2$. They are often described with the Gaussian distributions,

$$I(t) = I_0 \exp(-2t^2/\tau^2) \tag{2.4}$$

where I_0 is the peak intensity and the pulse duration is characterized by τ. Because of the Fourier transform relation between $U(t)$ and $V(t)$, the temporal width is inversely proportional to the spectral width, and related by $\tau = \sqrt{2\ln 2}/\pi\Delta\nu$, where $\Delta\nu$ is the bandwidth at full width half maximum (FWHM). Therefore, for a transform limited pulse (pulse without a chirp), the shortest pulse duration is set by the bandwidth of the laser pulse. When there are higher order terms in the phase, the pulse can be chirped and the intensity profile can deviate from a Gaussian distribution. By adjusting the grating spacing in the compressor, laser pulse duration, chirp and shape can be altered, and different physics can be investigated in LPA experiments.

2.2.2 TREX Laser System

The laser system used in this thesis experiment is a 40 TW CPA based laser called TREX at the LOASIS facility, LBNL. The layout of the TREX laser system is shown in Fig. 2.2. The laser pulses generated in the oscillator are amplified through five amplifying stages. Before the final amplification at the Main amplifier, the pulses are ~250 ps long, ~50 mJ/pulse. After the main amplifier, the energy of the

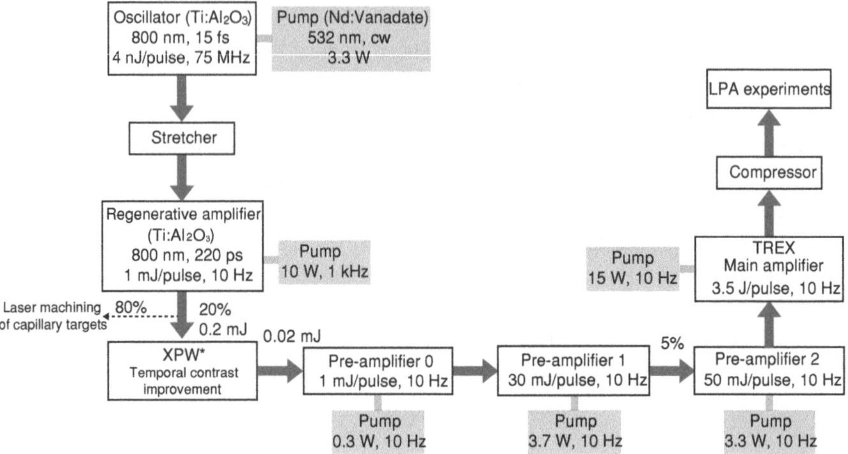

Fig. 2.2 Layout of TREX laser system at the LOASIS facility. Laser pulse duration, energy, and repetition rates are indicated. *Cross-polarized wave (XPW) system enhances temporal contrast of laser pulses [31]

laser is ~3.5 J/pulse, of which ~1.7 J/pulse is transported to LPA experiments. A cross-polarized wave system, labelled as XPW, enhances temporal contrast (intensity ratio between the pulse peak and the background) of laser pulses through wave generation in a nonlinear crystal [31]. A high contrast ratio of $\sim 10^8$ to 10^9 is critical in LPA experiments, especially when the pedestal of the laser pulse is expected to ionize gas. To minimize focus shift due to thermal lensing, the TREX main amplifier is cryogenically cooled. The pulses are temporally compressed to ~40 fs and spatially focused to ~20 μm to achieve the peak intensity of $\sim 10^{18}\ \mathrm{W/cm^2}$. In the staging experiment, each laser pulse is split into two pulses to drive the two stages. Details of the experiments will be discussed in Chap. 3.

2.3 Theory of Laser Propagation

2.3.1 Laser Diffraction

Laser diffraction imposes a limit to the energy gain in an LPA if not controlled. In this section, a mathematical formalism of laser diffraction is introduced. The optical wave is represented as a complex function of the form,

$$U(\mathbf{r}, t) = U(\mathbf{r}) \exp(i\omega t). \tag{2.5}$$

The complex amplitude takes the form $U(\mathbf{r}) = A(\mathbf{r}) \exp(i\phi(\mathbf{r}))$, where $A(\mathbf{r})$ is the amplitude and $\phi(\mathbf{r})$ is the phase of the wave at a given position $\mathbf{r}$. This wavefunction satisfies the wave equation,

$$\nabla^2 U - \frac{1}{c_0^2}\frac{\partial^2 U}{\partial t^2} = 0. \tag{2.6}$$

An optical wave is experimentally characterized with an intensity distribution $|U(\mathbf{r}, t)|^2$ and a wavefront which is a surface of constant phase $\phi(\mathbf{r})$.

An example of a paraxial solution to the wave equation (2.6) is a Gaussian pulse and uses Rayleigh length, z_R, as the characteristic length for diffraction. The waves are considered to be paraxial when the wavefronts are normal to paraxial rays. Paraxial waves imply that the envelope of the wave varies slowly with respect to its wavelength, $\lambda = 2\pi/k$, so that the complex amplitude can be described as, $U(r, z) = A(r, z)\exp(-ikz)$, where z is the propagation distance. Then, the complex amplitude of a Gaussian pulse is represented as,

$$U(r, z) = A_0 \frac{r_0}{r_s(z)} \exp\left(-\frac{r^2}{r_s^2(z)}\right) \exp\left(-ikz - ik\frac{r^2}{2R_c(z)} + i\zeta(z)\right), \tag{2.7}$$

where $r_s(z)$ is the laser spot size at z, r_0 is the laser spot size at focus ($z = 0$), $R_c(z)$ is the radius of curvature, and $\zeta(z)$ is the Guoy phase shift [32]. These parameters are related to z_R by,

$$\begin{aligned} z_R &= \frac{\pi r_0^2}{\lambda} \\ r_s(z) &= r_0\left[1 + (z/z_R)^2\right]^{1/2} \\ R_c(z) &= z\left[1 + (z_R/z)^2\right] \\ \zeta(z) &= \arctan\frac{z}{z_R}. \end{aligned} \tag{2.8}$$

Other paraxial solutions to the wave equation (2.6) include Laguerre-Gaussian and Hermite-Gaussian pulses and detailed descriptions of these pulses can be found elsewhere [32, 33]. In this thesis, the effects of a transverse laser profile on wake excitation in an LPA are investigated by characterizing the drive laser with a Gaussian and a higher order Laguerre-Gaussian pulse. This analysis will be presented in Chap. 6. The propagation properties of the Gaussian pulse in vacuum are determined by r_0 and λ, which define z_R. Figure 2.3 shows on-axis intensity and laser pulse radius as a function of z_R which is the distance in which the intensity decreases by half from the focus and the pulse radius increases by $\sqrt{2}$. Due to this reduction in intensity, the acceleration length in an LPA is limited to a few z_R if diffraction is not compensated.

In experiments, the laser intensity profile is described by a spot size at focus r_0 and a Strehl ratio (SR) in the Gaussian pulse framework. The radius of a Gaussian intensity distribution is defined by

$$I(r) = I_0\, e^{-2r^2/r_0^2}. \tag{2.9}$$

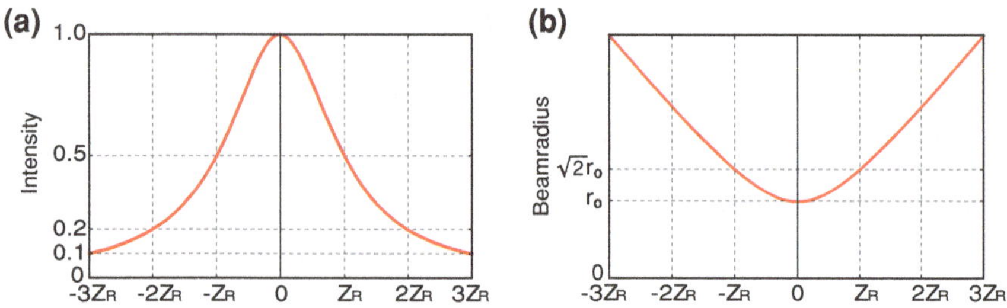

Fig. 2.3 Gaussian pulse on-axis intensity (**a**) and spot size (**b**) as a function of z_R. The graphs are normalized to the values at the focus

The SR is a measure of quality of the pulse profile compared to a Gaussian profile. It is defined as SR$= I_{\text{peak}}/I_0$, the ratio of the measured intensity peak I_{peak} to the peak intensity of the Guassian profile I_0. For TREX pulses used in the experiments discussed in this thesis, $r_0 \simeq 20\,\mu\text{m}$, SR$\sim$ 0.7–0.9, and the intensity is reduced by an order of magnitude within 5 mm (3 z_R).

2.3.2 Laser Guiding in Plasma Channel

Laser guiding is used to compensate for diffraction and to retain high laser intensity to extend acceleration lengths in LPAs. A plasma channel with a parabolic density profile can provide a focusing force that compensates for the diffraction of a Gaussian pulse. The mathematical formalism of a Gaussian laser propagating through a channel is as follows. Consider a parabolic plasma density profile [34, 35],

$$n(r) = n_0 + \Delta n r^2/r_{\text{m}}^2, \tag{2.10}$$

where r is the radial position, n_0 is the on-axis density, and Δn is the channel depth at a matched spot size, r_{m}. The index of refraction for this channel is,

$$\eta_r = 1 - \frac{\omega_{\text{p}}^2}{2\omega_{\text{L}}^2}\left(1 + \frac{\Delta n r^2}{n_0 r_{\text{m}}^2}\right). \tag{2.11}$$

When $\Delta n = \Delta n_c = (\pi r_{\text{m}}^2 r_e)^{-1}$, where $r_e = e^2/m_e c^2$ is the classical electron radius, the channel can provide guiding for a laser pulse with a Gaussian intensity profile, $|a|^2 = (a_0 r_0/r_{\text{s}})^2 \exp(-2\, r^2/r_{\text{s}}^2)$. In this equation, a_0, referred to as the normalized laser vector potential, is given by

$$a_0^2 \simeq 7.3 \times 10^{-19}\, (\lambda[\mu\text{m}])^2 I_0[\text{W/cm}^2]. \tag{2.12}$$

Throughout this thesis, a_0 is used to specify the strength of a laser pulse. Analysis of the paraxial equation shows that the laser evolves in the plasma channel as [5],

$$\frac{\mathrm{d}^2 R}{\mathrm{d}z^2} = \frac{1}{Z_M^2 R^3}\left(1 - \frac{\Delta n}{\Delta n_c} R^4\right), \tag{2.13}$$

where $R = r_\mathrm{s}/r_\mathrm{m}$ and $Z_\mathrm{M} = \pi r_\mathrm{m}^2/\lambda$. The first term on the right-hand side of Eq. (2.13) represents vacuum diffraction and the second term represents the focusing effects of the plasma channel. The general solution to Eq. (2.13) with the initial condition $r_\mathrm{i} = r_0$ and $\mathrm{d}r_\mathrm{s}/\mathrm{d}z = 0$ (i.e. laser focused at the entrance of the plasma channel) is,

$$r_\mathrm{s}^2 = \frac{r_\mathrm{i}^2}{2}\left[1 + \frac{r_\mathrm{m}^4}{r_\mathrm{i}^4} + \left(1 - \frac{r_\mathrm{m}^4}{r_\mathrm{i}^4}\right)\cos\left(\frac{2\lambda z}{\pi r_\mathrm{m}^2}\right)\right], \tag{2.14}$$

where r_i is the spot size at the entrance of channel, and z is the propagation distance [35]. Matched guiding ($r_\mathrm{s} = r_\mathrm{i}$) is achieved for a low power and low intensity ($a_0^2 \ll 1$) pulse when r_0 equals r_m and the laser is focused at the entrance of the channel. Here, low power is defined as the power below the critical power for the relativistic self-focusing effects, $P \ll P_c = 2c(e/r_\mathrm{e})^2(\omega_\mathrm{L}/\omega_\mathrm{p})^2$, which will be described in detail in Sect. 2.3.3. Laser spot size evolution of matched and mismatched guiding at low intensity and low power are calculated using Eq. (2.13) and are shown in Fig. 2.4a. The entrance of the capillary is set at $z = 0$, and the plasma channel is indicated as the cyan rectangle. For matched guiding, $r_0 = r_\mathrm{m} = 18\,\mu\mathrm{m}$, the laser retains its spot size as it propagates through the channel. For experiments

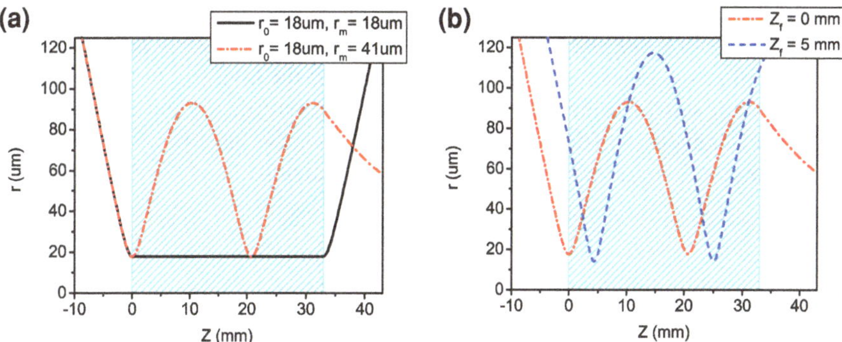

Fig. 2.4 Calculated laser spot size as a function of propagation distance in plasma channels which are indicated as shaded regions. **a** Matched guiding (*solid black*) for the case of $r_0 = r_\mathrm{m} = 18\,\mu\mathrm{m}$ and mismatched guiding (*dot-dashed red*) for the case of $r_0 = 18\,\mu\mathrm{m}$ and $r_\mathrm{m} = 41\,\mu\mathrm{m}$. Laser is focused at the entrance of plasma channel, $z = 0\,\mathrm{mm}$. **b** Evolution of spot sizes when pulses are focused at different locations of the same channel. Laser is focused at $z = 0\,\mathrm{mm}$ (*dot-dashed red*) and $z = 5\,\mathrm{mm}$ (*dashed blue*) for mismatched guiding with $r_0 = 18\,\mu\mathrm{m}$ and $r_\mathrm{m} = 41\,\mu\mathrm{m}$

discussed in this thesis, the laser underwent mismatched guiding. One of the guiding conditions used in the experiment was $r_0 = 18\,\mu\text{m}$ and $r_\text{m} = 41\,\mu\text{m}$, and shown as the dot-dashed red curve. This mismatched guiding leads to a significant oscillation of the spot size. The guiding condition also changes when the laser is focused at different locations with respect to the channel entrance. Figure 2.4b shows the laser evolution for cases where the laser is focused at $z = 0\,\text{mm}$ and $z = 5\,\text{mm}$. Changing the laser focus position changes the divergence of the laser at the entrance of the channel, resulting in a different spot size evolution. Even with this intensity oscillation, the laser intensity is maintained higher than it would in a vacuum. By extending the acceleration length in an LPA using a plasma channel, e-beams were accelerated to 1 GeV in 3 cm, demonstrating an average accelerating field >30 GV/m in 2006 [15].

2.3.3 *Relativistic Self-focusing*

Relativistic self-focusing can focus the laser more tightly in a plasma channel than it would in a vacuum. A heuristic picture of self-focusing is introduced in this section. A more complete derivation can be found in Ref. [12]. When the laser intensity is high, the increase in the effective mass of the electrons changes the plasma frequency to $\omega_\text{p}'^2 = \omega_\text{p}^2/\gamma$. This is expressed in the index of refraction by

$$\eta(r) \simeq 1 - \frac{\omega_\text{p}^2}{\omega_0^2}\frac{n_e(r)}{n_0\gamma(r)}, \tag{2.15}$$

where ω_p is the frequency of on-axis plasma density n_0, and $n_e(r)$ is the radial distribution of electron density, and $\gamma(r)$ is the relativistic factor associated with the transverse electron motion. Since $\gamma^2 \simeq (1 + |a|^2/2)^{1/2}$ for linearly polarized light, the index of refraction is modified as to further focus the laser to a higher a_0.

Relativistic self-focusing allows for power and intensity dependent laser evolution in plasma channels. Self-focusing is triggered when the laser power exceeds the critical power, $P_c = 2c(e/r_\text{e})^2(\omega_\text{L}/\omega_\text{p})^2$. In practical units,

$$P_c[\text{GW}] \simeq 17.5\left(\frac{\omega_\text{L}}{\omega_\text{p}}\right)^2. \tag{2.16}$$

When $P \gg P_c$, the laser spot evolution in a plasma channel [Eq. (2.13)] is modified to

$$\frac{\text{d}^2R}{\text{d}z^2} = \frac{1}{Z_M^2 R^3}\left(1 - \frac{\Delta n}{\Delta n_c}R^4 - \frac{P}{P_c}\right). \tag{2.17}$$

Simulated spot size evolutions for low power and high power are shown in Fig. 2.5. The simulation framework, INF&NO, used for this study is described in Sect. 2.4 and

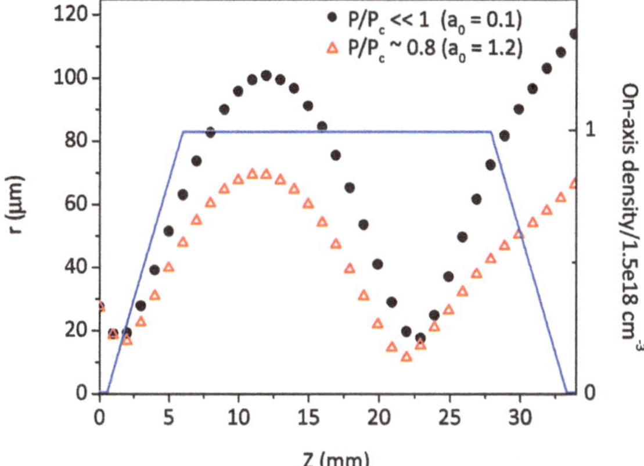

Fig. 2.5 Simulation showing the relativistic self-focusing effects in a plasma channel. Laser spot sizes r_s as a function of propagation distance for low power ($P/P_c \ll 1$) and high power ($P/P_c \simeq 0.8$) conditions used in experiments are shown. The on-axis plasma density is also shown with the blue line where $n_0 = 1.5 \times 10^{18}\,\text{cm}^{-3}$

details can be found in Ref. [36]. Laser spot sizes r_s for low power ($P/P_c \ll 1$) are shown as black dots and high power ($P/P_c = 0.8$) as red triangles as a function of propagation distance. The normalized plasma density is also shown. The simulations were performed for $r_0 = 18\,\mu\text{m}$, $r_m = 41\,\mu\text{m}$, $n_0 = 1.5 \times 10^{18}\,\text{cm}^{-3}$ and focused 1 mm into the channel as was performed in an experiment. The laser spot size for a high power laser in the channel is smaller than that for the low power laser due to self-focusing. In addition, the trailing edge of the laser pulse often exhibits stronger self-focusing than the leading edge of the pulse. This is because the index of refraction of the plasma is modified on the plasma frequency time scale and not the laser frequency time scale [5]. A detailed study of matched laser propagation in a plasma channel including the self-focusing effect is beyond the scope of this thesis and can be found elsewhere [37]. This numerical study shows that self-focusing causes power and intensity dependent laser evolution in a plasma channel, and can be used to achieve higher a_0 in plasma than in vacuum.

2.4 Plasma Waves

2.4.1 Plasma Wave Excitation

In LPAs, an underdense plasma is the medium to transfer the laser energy to accelerating electrons. Plasma is energetically the fourth state of the matter following solid, liquid, and gas in which the particles are ionized. The characteristic length of a plasma to screen a charge is known as the Debye length,

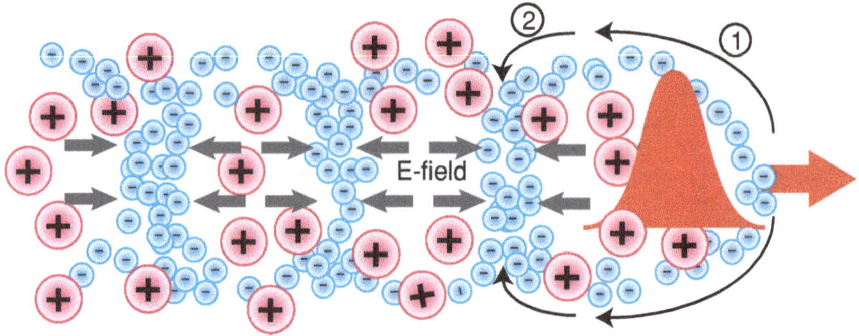

Fig. 2.6 A schematic of plasma wave excitation by an intense laser pulse. The radiation pressure of the drive laser pushes electrons out. Ions are considered to be stationary. Then, positive ions pull the electrons back, initiating electron density oscillations near the plasma frequency

$$\lambda_D = \sqrt{\frac{\epsilon_0 k_B T_e}{n_e e^2}}, \tag{2.18}$$

where ϵ_0 is the permittivity of free space, k_B the Boltzmann constant, e the electron charge, T_e the electron temperature, and n_e the electron density. Since ions move much more slowly than electrons, the ion term is neglected. For typical plasmas discussed in this thesis, $n_e \sim 10^{18}\,\text{cm}^{-3}$ and $T_e \sim 10\,\text{eV}$, resulting in $\lambda_D \sim 24\,\text{nm}$. Motions less than λ_D can be neglected in the laser-plasma interaction.

Plasma waves (wakefields) are driven by the radiation pressure (also known as the ponderomotive force) of the driving laser field. A conceptual picture of wake excitation is discussed at first, and a mathematical framework will be presented afterwards. A schematic is shown in Fig. 2.6. The gradient of the laser intensity, the ponderomotive force, pushes electrons away from the propagation axis and creates a charge separation between the electrons and ions in the plasma (① in Fig. 2.6). Since the ions are much more massive than the electrons, the ions are considered stationary. After the laser pulse has passed, the restoring force described by Gauss's law initiates a local density oscillation with a characteristic plasma frequency of $\omega_\text{p} = 2\pi c/\lambda_\text{p}$ (② in Fig. 2.6). The plasma frequency is related to the background electron density, n_0, by

$$\omega_\text{p} = \sqrt{\frac{4\pi e^2 n_0}{m_\text{e}}}, \tag{2.19}$$

where the m_e is the electron rest mass. The associated plasma wavelength in practical units is given by,

$$\lambda_\text{p}(\mu\text{m}) \cong 3.3 \times 10^{10}/\sqrt{n_0(\text{cm}^{-3})}. \tag{2.20}$$

This plasma wavelength is the characteristic scale for the accelerating structures in LPAs. For $n_0 \sim 10^{18}\,\text{cm}^{-3}$, $\lambda_p \sim 33\,\mu\text{m}$ and the λ_p is longer for a lower density and shorter for a higher density. The laser pulse propagates through the plasma at a group velocity, $v_g = c(1-\omega_p^2/\omega_L^2)^{1/2}$, and continues to initiate the electro-static charge density oscillation. The phase velocity of the plasma wave roughly corresponds to the group velocity of the driving laser, $v_p \sim v_g$. This process is the excitation of a plasma wave and provides the accelerating structure to charged particles.

The amplitude of the excited field depends on plasma density n_0 and laser strength a_0. Ionized plasmas can sustain a plasma wave with electric fields in excess of $E_0 = c\, m_e\, \omega_p/e$, where c is the speed of light in vacuum. This is known as the cold nonrelativistic wave breaking field [5]. In practical units,

$$E_0(\text{V/m}) \cong 96\sqrt{n_0(\text{cm}^{-3})}. \tag{2.21}$$

This field corresponds to ~96 GV/mfor an electron density of $n_0 \sim 10^{18}\,\text{cm}^{-3}$. For example, using a linearly polarized square pulse of the optimized pulse length, the maximum excited field in the one-dimensional (1-D) limit is,

$$E_{\max}(n_0, a_0) = E_0(n_0)\,\frac{a_0^2/2}{\sqrt{1+a_0^2/2}}. \tag{2.22}$$

Figure 2.7 shows $E_{\max}$ for several values of a_0 as a function of n_0. It is possible for the maximum amplitude of a nonlinear plasma wave to exceed E_0. The nonlinear, relativistic, cold fluid wave breaking field is,

$$E_{\text{WB}} = \sqrt{2}(\gamma_p - 1)^{1/2} E_0. \tag{2.23}$$

When the field exceeds the wave breaking limit, the velocity of the background plasma electrons can exceed the phase velocity of the plasma wave and be trapped and accelerated in the plasma wave. For the parameters accessible in the experimental

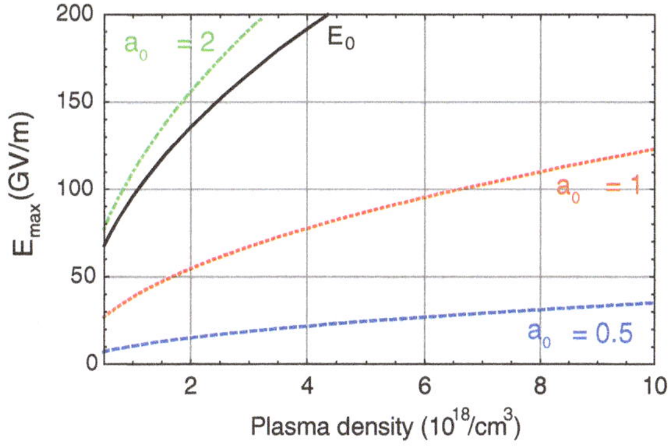

Fig. 2.7 Maximum electric fields $E_{\max}$ in 1-D limit calculated using Eq. (2.22). E_0 is the cold nonrelativistic wavebreaking field defined by Eq. (2.21)

system discussed in this thesis ($n_0 \sim 1$–$5 \times 10^{18}\,\text{cm}^{-3}$, $a_0 \sim 0.1$–2), the E_{max} ranges between 1–20 GV/m. In addition, to excite wakefields of a given amplitude at lower n_0 requires a higher a_0 laser. For example, $E_{\text{max}} \sim 80\,\text{GV/m}$ is achieved with $n_0 \sim 0.5 \times 10^{18}\,\text{cm}^{-3}$ and a_0 of 2 while the same amplitude can be excited with a_0 of 1 if operated at $n_0 \sim 4 \times 10^{18}\,\text{cm}^{-3}$. The balance between n_0 and a_0 influences other effects such as electron dephasing and pump depletion which will be discussed later in this chapter.

The mathematical description follows the formalism presented in Refs. [5, 38]. Introducing the vector potential $\mathbf{A}$ and the electro-static potential Φ, the electric and magnetic fields can be expressed as $\mathbf{E} = -\nabla\Phi - \partial\mathbf{A}/\partial ct$ and $\mathbf{B} = \nabla \times \mathbf{A}$. In the Coulomb gauge ($\nabla \cdot \mathbf{A} = 0$), $\mathbf{A}$ describes the laser and the Φ describes the plasma wave. The potentials are generally normalized as $|\mathbf{a}| = e|\mathbf{A}|/m_e c^2$ and $\phi = e\Phi/m_e c^2$. For a given laser pulse, the amplitude of the laser vector potential a_0 as defined in Eq. (2.12) is related to the transverse laser electric field by $E_L[\text{TVm}^{-1}] = 3.2\,a_0/\lambda[\mu\text{m}]$.

The excited plasma wave can be described by the density distribution n, the electric field E_z, or the potential ϕ. These values are related through $\mathbf{E} = -\nabla\phi$ and Gauss's law, $\nabla \cdot \mathbf{E} = -4\pi e(n - n_0)$. The wave equation for the density perturbation by the ponderomotive force is,

$$\left(\frac{\partial^2}{\partial t^2} + \omega_p^2\right)\frac{\delta n}{n_0} = c^2\nabla^2\frac{a^2}{2}, \tag{2.24}$$

where $\delta n/n_0 = (n - n_0)/n_0$ and $\mathbf{F}_p \propto c^2\nabla(a^2/2)$ is the ponderomotive force. A solution to the wave equation in the linear regime ($a \ll 1$) is,

$$\delta n/n_0 = (c^2/\omega_p)\int_0^t dt' \sin\left[\omega_p(t - t')\right]\nabla^2 a^2(\mathbf{r}, t')/2. \tag{2.25}$$

The sinusoidal form illustrates that the wakefield excitation is most efficient when the laser envelope length characterizing the extent of a^2 is on the order of λ_p.

In order to understand the wake excitation in the nonlinear regime ($a \gg 1$), a 1-D limit (a large transverse laser spot size) is considered. Assuming the laser pulse is slowly evolving and the phase velocity of the wave is near the speed of light, $\gamma_p^2 \gg 1$ where $\gamma_p = (1 - v_p^2/c^2)^{-1/2}$, the wake potential can be described by [10],

$$k_p^{-2}\frac{\partial^2\phi}{\partial\zeta^2} = \frac{(1 + a^2)}{2(1 + \phi)^2} - \frac{1}{2}. \tag{2.26}$$

This equation is valid in general for all a. Linear and nonlinear plasma densities and electric fields excited by laser pulses with a half-sine intensity profile are shown in Fig. 2.8. A linear regime plasma wave ($a_0 = 0.2$) is shown in Fig. 2.8a and a nonlinear regime plasma wave ($a_0 = 2$) is shown in (b). In the linear regime, the plasma wave is sinusoidal and the plasma wavelength is expressed by Eq. (2.20). In the nonlinear

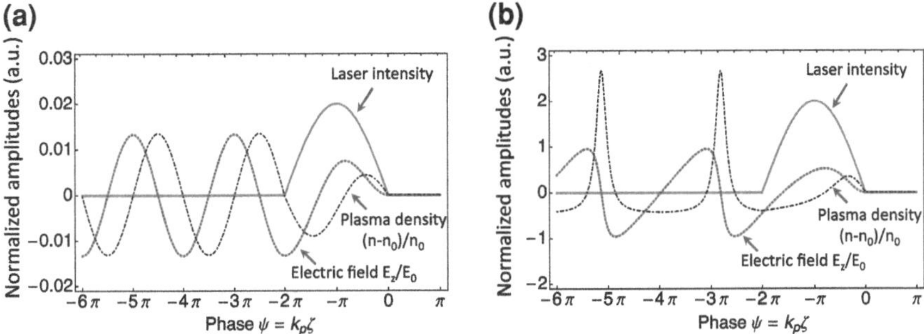

Fig. 2.8 Plasma waves and electric fields excited by laser pulses of half-sine intensities in 1-D limit. The wake potential is calculated by Eq. (2.26). **a** Linear regime where $a_0 = 0.2$. **b** Nonlinear regime where $a_0 = 2$

regime, the density profile steepens and the electric field becomes a sawtooth shape. The plasma wavelength in the nonlinear regime $\lambda_{\rm NP}$ is larger than that in the linear regime and depends on the wake amplitude:

$$\lambda_{\rm NP} = \lambda_{\rm p} \begin{cases} 1, & \text{for } E_{\rm max}/E_0 \ll 1 \\ (2/\pi) E_{\rm max}/E_0, & \text{for } E_{\rm max}/E_0 \gg 1. \end{cases} \tag{2.27}$$

This 1-D illustrations show the different plasma wave shapes and wavelengths between linear and nonlinear regimes.

Numerical investigations are required to understand both transverse and longitudinal electric fields in three-dimensions (3-D). Simulations using the INF&NO framework were used to demonstrate wake excitation in linear and nonlinear regimes. INF&NO is a two-dimensional cylindrical $(r - z)$ particle-in-cell (PIC) code that adopts an envelope model for the laser pulse [36]. Laser-plasma interactions are described using the ponderomotive-force approximation [5]. Figure 2.9 shows simulated plasma density, E_z and E_r in linear and nonlinear regimes. Normalized plasma density distribution in a linear regime with $a_0 \sim 0.5$ and $n_0 = 1.5 \times 10^{18}\,\text{cm}^{-3}$ is shown in Fig. 2.9a. In the linear regime, accelerating and decelerating fields are symmetric in size and shape as shown in Fig. 2.9b. Similarly, transverse focusing and defocusing fields are symmetric as shown in Fig. 2.9c. As a result, there is approximately $\lambda_{\rm p}/4$ phase region where electrons can be both accelerated and focused as indicated with dashed red lines. In the nonlinear regime, the laser pulse almost completely blows out the plasma electrons near the back of the pulse as shown in Fig. 2.9d for $a_0 \sim 2$ and $n_0 = 1.5 \times 10^{18}\,\text{cm}^{-3}$. The high electron density region outlines a shape of a bubble. This region is also called the bubble or blowout regime. The accelerating and focusing regions for electrons (bounded by dashed red lines) are larger than those in the linear regime Fig. 2.9e and f. However, for positively charged particles, the focusing region is much smaller in the nonlinear regime than in the linear regime. These simulations illustrate the difference in longitudinal and transverse wakefield profiles in linear versus nonlinear regimes.

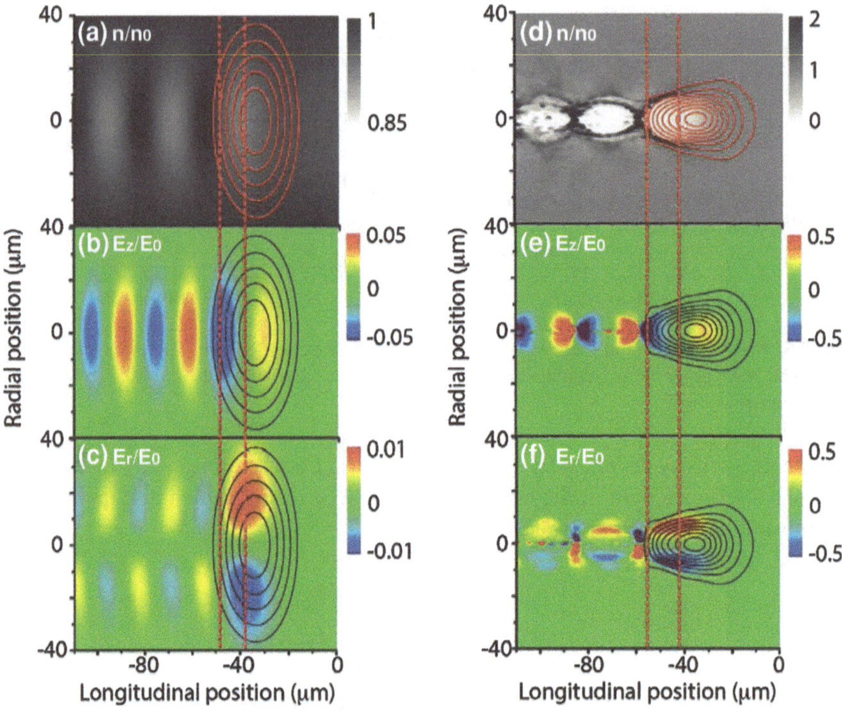

Fig. 2.9 PIC simulations of plasma wave excitations in linear (**a–c**) and nonlinear (**d–f**) regimes. The $a_0 \sim 0.5$ for the linear regime and $a_0 \sim 2$ for the nonlinear regime with $n_0 = 1.5 \times 10^{18}\,\text{cm}^{-3}$. The laser intensity contours are indicated with *solid lines*. The *red dotted lines* indicate accelerating and focusing regions for electrons. The plasma density is normalized to n_0 and electric fields are normalized to $E_0 = 117\,\text{GV/m}$

In LPAs, the magnitudes of the transverse forces can also be extremely large, on the order of GV/m, requiring solid understanding and control for successful electron acceleration. In the linear to quasi-linear regime ($a_0 \lesssim 1$), the electric fields of the wake are proportional to the intensity profile. The longitudinal field is $E_z \sim \partial a^2/\partial z$ and the transverse field is $E_\perp \sim \nabla_\perp a^2$. Therefore, the focusing force can be manipulated by tailoring the transverse laser intensity profile. Section 6.2 will present an effort to study wakefield profiles excited by different transverse laser intensity profiles. In the blowout regime ($a_0^2 \gg 1$), the focusing field can be approximately the same order of magnitude as the accelerating field, $E_z/E_0 \sim k_\text{p}\zeta/2$ and $(E_r - B_\theta)/E_0 = (k_\text{p}r/2)$ where E_r is the radial electric field and B_θ is the azimuthal magnetic field [5, 39–41]. This can be observed in simulated E_r/E_0 shown in Fig. 2.9f, which are roughly the same magnitudes as E_z/E_0 shown in Fig. 2.9e. At the edge of an e-beam with the spot size σ_r, the field magnitude is estimated as

$$E_r(\text{GV/m}) \simeq 9 \times 10^{-18}\, n_0(\text{cm}^{-3})\sigma_r(\mu\text{m}). \tag{2.28}$$

For an LPA with $n_0 \sim 10^{18}\,\text{cm}^{-3}$ and $\sigma_r \sim 0.1\,\mu\text{m}$, $E_r \sim 1\,\text{GV/m}$. Such a large focusing force implies the necessity of extremely precise control over the forces the electrons experience.

In summary, the basic physics of wake excitation by an intense laser pulse was introduced. In the linear regime ($a_0^2 \ll 1$), the plasma wave is sinusoidal and of the length $\lambda_p \propto n_0^{-1/2}$. In the nonlinear regime ($a_0^2 \gg 1$), the plasma wave steepens and the plasma wavelength increases. The amplitude of the maximum accelerating field, E_{max}, is a function of plasma density, laser strength and oscillations of focus spot size in waveguides. Large transverse forces require precise control over the electric fields and accelerating electrons in LPAs.

2.4.2 Electron Acceleration and Dephasing

In 1-D limit, electron trapping, acceleration, and dephasing can be studied by considering electron momentum and phase with respect to the plasma wave. The motion of an electron is described by a constant Hamiltonian [5],

$$H(\gamma, \psi) = \gamma(1 - \beta\beta_p) - \phi(\psi). \tag{2.29}$$

Figure 2.10 shows the electron trajectory in momentum-phase space for a constant Hamiltonian. Here, we consider a sinusoidal plasma wave, $\phi(\psi) = \phi_0 \cos(\phi)$, where $\phi_0 = 0.01$ and the phase velocity of the plasma wave $\gamma_p = 20$, and the frame is moving at the phase velocity. Trajectories of background electrons are indicated as

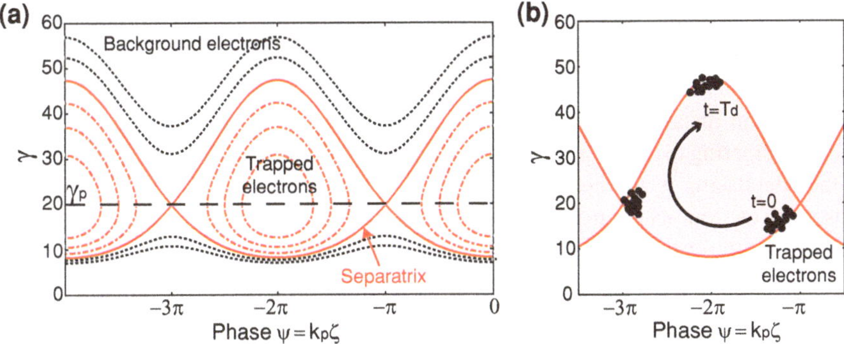

Fig. 2.10 Electron trajectory in momentum-phase space for a constant Hamiltonian. A sinusoidal plasma wave with $\gamma_p = 20$ and $\phi_0 = 0.01$ was used. **a** Trajectories of background electrons that have too much or too little momentum are shown as *black dotted curves*. These electrons are not trapped in the plasma waves. Trajectories of trapped electrons are shown as *red curves*. **b** Acceleration of trapped electrons illustrated in the momentum-phase space. Initially low energy electrons $t = 0$ outruns the plasma wave at time $t = T_{\text{d}}$

dotted black curves. Electrons on these trajectories have either too much or too little initial momentum, so they slip out of the plasma waves. These electrons experience accelerating and decelerating fields, but they will not be trapped in the wake. If an electron has enough initial momentum to keep up with the wake, the electron can be trapped in the plasma wave. The solid red curves indicate the momentum-phase trajectory of a trapped electron, $H = H(\gamma_p, \psi_T)$. This trajectory is referred to as the separatrix.

When the accelerating electrons gain enough energy, they will become relativistic and outrun the plasma wave. Figure 2.10b illustrates the acceleration of trapped electrons from $t = 0$ to $t = T_{\mathrm{d}}$. The electrons at $t = T_{\mathrm{d}}$ are now traveling at or greater than the phase velocity of the plasma wave and can no longer gain energy. This is referred to as dephasing, and is one of the limitations to the electron energy gain in LPAs. The characteristic distance of the dephasing length L_{d} is defined by the distance in which a highly relativistic electron phase slips by $\lambda_{\mathrm{p}}/4$, roughly the phase region that is both focusing and accelerating. The dephasing lengths in linear and nonlinear regimes are given by [5],

$$L_{\mathrm{d}} \simeq \frac{\lambda_{\mathrm{p}}^3}{2\lambda^2} \times \begin{cases} 1, & \text{for } a_0^2 \ll 1 \\ (\sqrt{2}/\pi)a_0/N_p, & \text{for } a_0^2 \gg 1, \end{cases} \tag{2.30}$$

where N_p is the number of plasma periods behind the laser pulse. The $1/N_p$ dependence in the nonlinear regime is from the plasma wave period increasing as the laser pulse steepens, which is the dominant effect in determining the plasma wave phase velocity. Consequently, the electron beam outruns the plasma wave faster (i.e. shorter L_{d}). For $n_0 \sim 10^{18}\,\mathrm{cm}^{-3}$, $L_{\mathrm{d}} \simeq 25$–$30\,\mathrm{mm}$ for $a_0 < 2$. This length is the motivation behind using 33 mm plasma channels in experiments discussed in this thesis.

Dephasing can be mitigated by operating at lower n_0 and/or appropriately tailoring the plasma density. Since $\lambda_{\mathrm{p}} \propto n_0^{-1/2}$, the accelerating electron can be kept at a fixed phase if λ_{p} changes to compensate for the phase slippage of the electrons due to acceleration. This can be achieved if the plasma density is increased with propagation distance, decreasing the plasma wavelength. More detailed theory on density tapering to compensate for the dephasing can be found elsewhere [29, 42, 43]. Since dephasing is another limitation to the acceleration length in LPAs following laser diffraction, its control is critical and future experimental investigations of the dephasing will be valuable.

2.4.3 Electron Beam Production

Controlled injection of electrons in a plasma wave is an active area of research in the LPA community. Recent progress includes experimental demonstration of controlled injection based on techniques using colliding pulses, ionization, and negative density gradients [20, 23, 24, 44, 45]. In this section, electron injection mechanisms using self-trapping, ionization of nitrogen atoms, and negative plasma density gradients will be discussed. All of these injection mechanisms were demonstrated using the TREX laser system and the results will be presented in Chap. 4.

2.4.3.1 Injection Through Self-trapping

Self-trapping refers to the e-beam injection and trapping of background plasma electrons by exciting large amplitude plasma waves. Electron beam generation via self-trapping in a capillary waveguide is experimentally more straightforward than other methods, but requires high a_0 to excite the large amplitude waves. The self-trapping mechanism is explained at first in 1-D limit to illustrate the concept. The adaptation to 3-D will be qualitatively described afterwards.

In the 1-D limit, background electrons are injected when the excited wake amplitude exceeds the wave breaking field. The wave breaking field $E_{\rm WB}$ is the maximum electric field the plasma can sustain. In the cold fluid model, the limit is characterized as the field if all electrons were oscillating at $\omega_{\rm p}$, and is expressed in Eq. (2.21). Where the plasma wave is travelling at a large phase velocity, $\gamma_{\rm p}$, the plasma can sustain $E_{\rm WB} > E_0$ where $E_{\rm WB}$ was introduced in Eq. (2.23). In other words, the electrons are injected if the longitudinal velocity of electrons exceeds the phase velocity of the plasma wave. Figure 2.11a is the schematic of self-trapping in 1-D and Fig. 2.10a is the electron momentum-phase space trajectory, indicating background and trapped electron trajectories. When the wake amplitude increases, the momentum of background plasma electrons increases. When the electron momentum is sufficiently large, the electrons can enter the trapped separatrix.

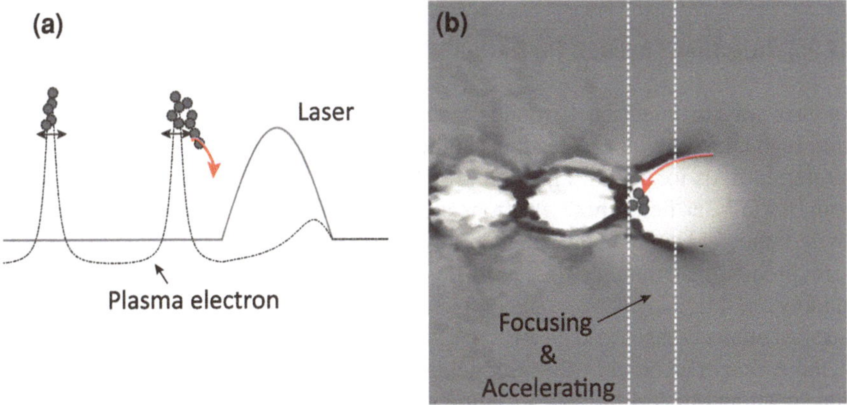

Fig. 2.11 Schematic of self-trapping of background plasma electrons in nonlinear plasma waves. **a** One-dimensional limit where background electrons are always injected via wave breaking. **b** In 3-D, laser expels all electrons behind the laser pulse in nonlinear regime. Some electrons traveling along the cavity wall, experience focusing and accelerating forces which place the electrons at the back of the bubble to be trapped

In 3-D, self-trapping is not always the longitudinal wave breaking phenomena [46]. As shown in Fig. 2.11b, increasing the laser intensity leads to a transition from sinusoidal waves to a bubble regime where all the electrons are pushed aside transversely and forms a cavity of plasma electrons. When the transverse momentum of the background electrons becomes high enough due to the ponderomotive force of the laser, the electrons can trace the wall of the plasma cavity (red arrow in the figure), experience the focusing force, and fall into the accelerating bucket. This mechanism can also lead to trapping of background electrons.

Thermal effects can lower this self-trapping threshold to below the wave breaking limit [5]. Background electron temperature is $T_e \approx 5$–$10\,\text{eV}$ for the plasma discussed in this thesis [47]. Due to the nonzero temperature of the plasma, the electrons have a momentum spread before the wake excitation. When the plasma wave is excited, the temperature of the electrons increases in the high density regions, $T_e \propto (n/n_0)^2$ [48]. The momentum of electrons at the density peaks of the plasma wave are further increased. This increase in electron temperature can trigger self-trapping before the wavebreaking limit.

Properties of e-beams depend on parameters such as laser evolution, beam loading and dephasing [8, 15]. Self-trapped monoenergetic e-beams with a few percent energy spread were demonstrated in 2004 by three groups [6–8]. Simulations suggest that fields from injected electron bunches reduced the plasma wakefield and terminated the injection in these experiments. The trapped electrons were then accelerated to the dephasing lengths which reduced the energy spread. When electron injection is not terminated or acceleration length is not optimized for dephasing, a broad energy spread e-beam can be produced. Electron bunch production via self-trapping was investigated for the staging experiment because of its simplicity in experimental setup. Experimental results will be presented in Sect. 4.2.

2.4.3.2 Injection Through Ionization of High *Z* Atoms

Another way to inject background electrons is to ionize deeply bound electrons from high atomic number (Z) atoms at the proper phase within the wakefield. This is referred to as ionization injection [45, 49]. When electrons are ionized near the peak of the laser field, these electrons are already inside the plasma wave and are easily trapped. This injection method lowers the threshold of laser intensity for e-beam production compared to the self-trapping injection when the high Z atoms are suitably chosen [24, 25]. A more complete analysis of the ionization injection using nitrogen atoms is discussed elsewhere [24, 25].

Ionization of an atom by the laser field is explained by two steps: Barrier suppression and tunneling ionization. Ionization via barrier suppression is realized when the laser field is large enough to remove bound electrons. The Coulomb potential of an atom is modified by the quasi-static laser field, $V(x) = -Ze/|x| - E_x x$ [50–52]. The first term is the Coulomb potential and the second term is the modification due to the laser field. A strong laser field above $E_L = U_i^2/(4eZ)$ where U_i is the ionization potential and Z is the charge state, lowers the potential below the ionization potential,

Table 2.1 Ionization potential, corresponding intensity and a_0 for hydrogen, helium and nitrogen atoms are listed

Gas	Charge state, Z	Ionization potential (eV)	Intensity (W/cm^2)	a_0
H	1	13.6	1.4×10^{14}	0.008
He	1	24.6	1.5×10^{15}	0.03
	2	54.4	8.8×10^{15}	0.06
N	1	14.5	1.8×10^{14}	0.009
	2	29.6	7.7×10^{14}	0.02
	3	47.5	2.3×10^{15}	0.03
	4	77.4	9.0×10^{15}	0.06
	5	97.9	1.5×10^{16}	0.08
	6	552.1	1.0×10^{19}	2.2
	7	667.1	1.6×10^{19}	2.7

To calculate a_0, $\lambda = 0.8\,\mu$m was assumed. Laser pulses of $\sim 10^{18}$ W/cm^2 will ionize H, He and the first five states of N atoms at the very leading edge of the pulse [53]

allowing the atomic electron to freely escape [50]. The required laser intensity for this barrier suppression ionization is

$$I[\mathrm{W/cm^2}] = 4 \times 10^9 (U_i^4[\mathrm{eV}]/Z^2). \tag{2.31}$$

In the experiments discussed in this thesis, nitrogen was used as the high Z atom and the neutral gas was balanced with helium. The ionization potential, ionization intensity and corresponding a_0 for $\lambda = 0.8\,\mu$m are listed in Table. 2.1. Since the peak intensities of laser pulses used in experiments are $\sim 10^{18}$ W/cm^2, hydrogen, helium and the first five ionization states of nitrogen are fully ionized at the leading edge of the drive pulse.

When the laser intensity is below the barrier suppression level, the ionization of an atom can occur through tunneling ionization. The probability of tunneling ionization can be calculated with the Keldysh model [51, 54, 55],

$$W(|E_\mathrm{L}|) = 4\left(\frac{3}{\pi}\right)^{1/2} \Omega_0 \left(\frac{U_i}{U_H}\right)^{7/4} \left(\frac{E_H}{|E_\mathrm{L}|}\right)^{1/2} \exp\left(-\frac{2}{3}\left(\frac{U_i}{U_H}\right)^{3/2} \frac{E_H}{|E_\mathrm{L}|}\right), \tag{2.32}$$

where $\Omega_0 = 4 \times 10^{16}\,\mathrm{s}^{-1}$ is the characteristic atomic frequency, $U_H = 13.6\,\mathrm{eV}$ is the ionization energy of hydrogen, and $E_H = 5.2\,\mathrm{GV/cm}$ is the ionization field of hydrogen, and E_L is the laser field.

Unlike the outer five electrons of nitrogen atoms that are ionized at the leading edge of the laser pulse, the inner sixth and seventh electrons are ionized near the peak of the laser field. Figure 2.12 shows the ionization probability for $N^{5+} \rightarrow N^{6+}$ calculated using Eq. (2.32). The black line is with $a_0 = 1.7$ and the blue dashed line

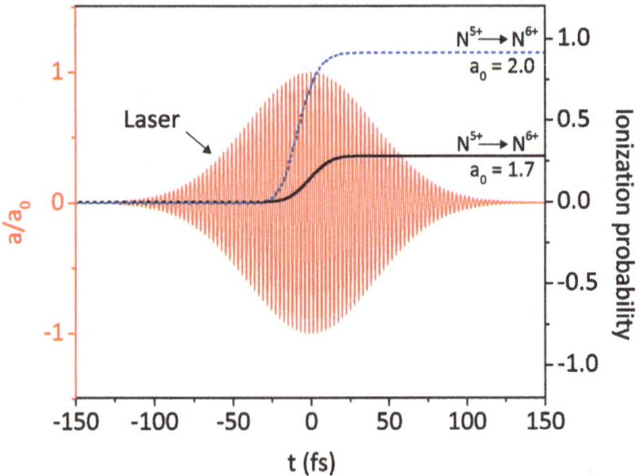

Fig. 2.12 Ionization probability for $N^{5+} \rightarrow N^{6+}$ and normalized laser vector potential, a/a_0, versus time within the laser pulse. The *black line* is for $a_0 = 1.7$ and the *blue dashed line* is for $a_0 = 2.0$. The laser $a_0 \sim 1.5$–2.0 is the threshold of ionization injection

is with $a_0 = 2.0$. The ionization probability is $< 30\,\%$ for $a_0 = 1.7$ and increases to $\sim 90\,\%$ for $a_0 = 2.0$. This illustrates that the sixth electron is only ionized near the peak of the laser pulse and $a_0 = 1.5$–2.0 is approximately the threshold of ionization injection using nitrogen atoms. When the laser focusing is assisted by effects such as relativistic self-focusing and pulse compression, the ionization injection can be facilitated. Suitability of this method to produce e-beams in the 1st module of the staging experiment was investigated and the results will be presented in Sect. 4.2.

2.4.3.3 Injection with Negative Density Gradient

Self-trapping injection can be facilitated by tailoring the plasma density. When a laser pulse propagates in a plasma with a negative density gradient ($dn/dz < 0$), the phase velocity at the back of the wake decreases as the laser propagates [19]. In Sect. 2.4, it was introduced that the plasma wavelength increases with decreasing density ($\lambda_p \propto n_0^{-1/2}$). Plasma wavelengths for varying plasma densities are illustrated in Fig. 2.13. When the laser pulse propagates in a region with decreasing density, the plasma wavelength increases ($\lambda_{p1} < \lambda_{p2}$ where $n_1 > n_2$). The density transition changes the location of the phase peak by $\Delta\zeta = \lambda_{p1} - \lambda_{p2}$, where $\zeta = z - ct$ is the distance behind the laser pulse. If this density transition occurs over a length L, the change in the phase velocity is $\Delta v_p \sim c\,\Delta\zeta/L$. As discussed earlier in this section, electron injection through self-trapping occurs when the velocity of the electrons v_e exceeds the phase velocity of the plasma wave. When the phase velocity of the back of the plasma wave is reduced, injection is facilitated [5]. This effect increases with the magnitude of the density gradient. Experimental results on this technique will also be presented in Sect. 4.2.

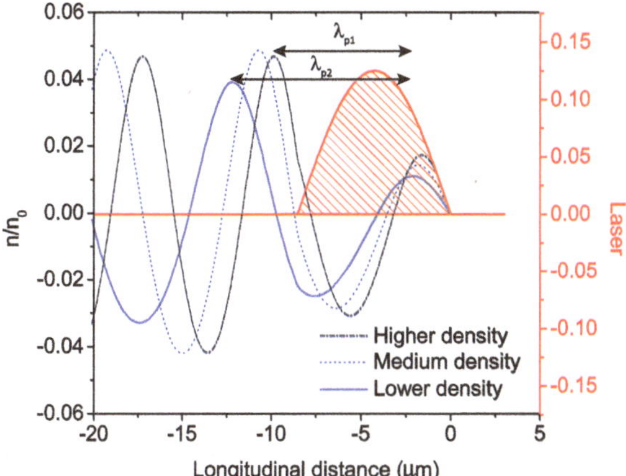

Fig. 2.13 Illustration of plasma wavelengths for three different plasma densities. When a laser propagates in a negative density gradient, phase velocity at the back of the plasma wave decreases due to the increasing λ_p and facilitates electron injection

2.5 Limitations to Energy Gain

2.5.1 Acceleration Limits

It was discussed in Chap. 1 that limitations to electron energy gain in a single LPA module are imposed by laser diffraction, electron dephasing and laser energy depletion. Laser diffraction and electron dephasing were introduced earlier in this chapter. In this section, energy depletion of the drive laser, also referred to as pump depletion will be introduced. These effects, diffraction, dephasing and depletion, limit the acceleration length and consequently possible energy gain a single stage LPA [5, 26].

As the laser excites the wakefield, the laser loses energy and eventually its intensity would be too small to further excite plasma waves. The characteristic length for the pump depletion is defined by the distance in which the laser pulse loses half of its energy to the plasma, $E_z^2 L_{\rm pd} \simeq E_L^2 L$ where E_L is the laser field. For a linearly polarized square pulse in the 1-D limit, this length is given by [5, 56],

$$L_{\rm pd} \simeq \frac{\lambda_{\rm p}^3}{\lambda^2} \times \begin{cases} 2/a_0^2, & \text{for } a_0^2 \ll 1 \\ (\sqrt{2}/\pi)a_0, & \text{for } a_0^2 \gg 1. \end{cases} \tag{2.33}$$

For n_0 of 10^{18} cm^{-3} and a_0 of 2, $L_{\rm pd}$ is greater than 50 mm. The pump deletion limit on acceleration length decreases when n_0 and a_0 are large.

A summary of the characteristic lengths for diffraction, dephasing and depletion is:

- Rayleigh length: $z_R = \pi r_0^2/\lambda$
  Length in which laser diffraction reduces the intensity by half (see Sect. 2.3).
- Dephasing length:
  $L_d \simeq \lambda_p^3/(2\lambda^2)$ for $a_0^2 \ll 1$ and $L_d \simeq (\lambda_p^3/2\lambda^2)(\sqrt{2}/\pi)a_0 N_p$ for $a_0^2 \gg 1$.
  Lengths in which electrons outrun the plasma wave by $\lambda_p/4$ to experience accelerating and focusing forces (see Sect. 2.4).
- Depletion length:
  $L_{pd} \simeq (\lambda_p^3/\lambda^2)2/a_0^2$ for $a_0^2 \ll 1$ and $L_{pd} \simeq (\lambda_p^3/\lambda^2)(\sqrt{2}/\pi)a_0$ for $a_0^2 \gg 1$.
  Lengths in which the laser loses half of its energy to the plasma through wake excitation.

The laser diffraction characterized by the Rayleigh length represents the shortest acceleration limit and has to be mitigated. For experiments discussed in this thesis, r_0 was 20 μm and z_R was 1.6 mm. Diffraction was mitigated by guiding the laser in a parabolic plasma channel which was discussed in Sect. 2.3.2.

In the linear regime ($a_0^2 \ll 1$), acceleration is limited by dephasing ($L_d \ll L_{pd}$). Dephasing and depletion lengths are plotted in Fig. 2.14. For example, when $a_0 = 0.5$, $\lambda = 0.8$ μm, and $n_0 = 10^{18}$ cm^{-3}, the dephasing length is $L_d \sim 3$ cm and the pump depletion length is $L_{pd} \sim 45$ cm. Methods to mitigate dephasing have been proposed, and experimental demonstrations are critical in the linear regime [29, 42, 43].

In the nonlinear regime ($a_0^2 \gg 1$), the dephasing length will increase and the pump depletion length will decrease, so that $L_d \sim L_{pd}$ as illustrated in Fig. 2.14. Dephasing

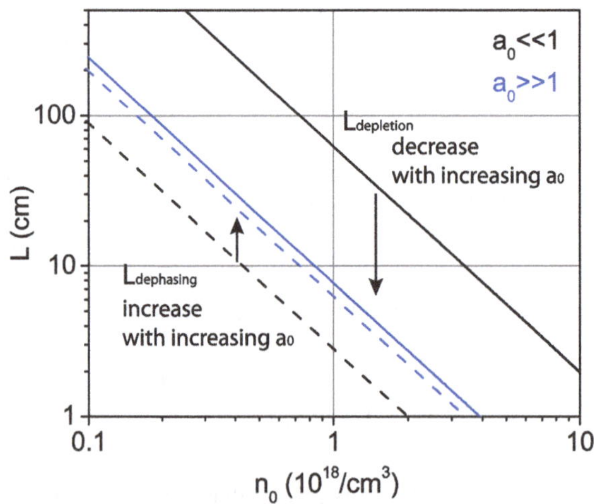

Fig. 2.14 Dephasing and depletion lengths as a function of plasma density for $a_0 \ll 1$ and $a_0 \gg 1$. For $a_0 \ll 1$, acceleration length is limited by dephasing. For $a_0 \gg 1$, both dephasing and depletion limits are similar

and depletion can be mitigated by decreasing the density since $L_{\mathrm{d}} \propto n_0^{-3/2}$ and $L_{\mathrm{pd}} \propto n_0^{-3/2}$. However, the decrease in plasma density requires a larger a_0 and a longer pulse for optimal wake excitation. In addition, the nonlinear regime also induces electron injection through self-trapping. For a dark-current free acceleration of externally injected e-beam, the nonlinear regime would not be suitable.

The quasi-linear regime characterized by $a_0^2 \sim 1$ and $L_{\mathrm{d}} \sim L_{\mathrm{pd}}$, is a favourable regime to optimize the energy gain. Laser energy is efficiently transferred to the plasma and externally injected e-beams can be accelerated without additional electron injection. Furthermore, positively charged particles can also be accelerated and focused. For these reasons, operation in the quasi-linear regime is considered to be ideal for acceleration stages in staged LPAs.

2.5.2 Scaling Laws for Energy Gain

The ideal energy gain in an LPA can be estimated by [5, 26],

$$\Delta W = eE_z L_{\mathrm{acc}}, \tag{2.34}$$

where L_{acc} is the acceleration length and E_z is the 1-D accelerating field given in Eq. (2.22). In practical units, if the acceleration length is limited by dephasing, $L_{\mathrm{acc}} \simeq L_{\mathrm{d}}$, the energy gain is,

$$W_{\mathrm{dp}}(\mathrm{MeV}) \simeq \frac{630\, I(\mathrm{W/cm^2})}{n_0(\mathrm{cm^{-3}})} \begin{cases} 1, & \text{for } a_0^2 \ll 1 \\ (2/\pi)/N_p & \text{for } a_0^2 \gg 1. \end{cases} \tag{2.35}$$

If the L_{acc} is limited by depletion, the ideal energy gain is,

$$W_{\mathrm{pd}}(\mathrm{MeV}) \simeq \begin{cases} 3.4 \times 10^{21}/(\lambda^2\,[\mu\mathrm{m}^2]\, n_0\,[\mathrm{cm^{-3}}]), & \text{for } a_0^2 \ll 1 \\ 400\, I\,[\mathrm{W/cm^2}]/n_0\,[\mathrm{cm^{-3}}] & \text{for } a_0^2 \gg 1. \end{cases} \tag{2.36}$$

In the quasi-linear regime, the maximum electron energy gain using $\lambda = 0.8\,\mu\mathrm{m}$ and $n_0 = 10^{18}\,\mathrm{cm^{-3}}$ when accelerated to the depletion limit is 5 GeV. These estimates are idealized and do not consider effects such as mismatched guiding, self-focusing or other laser-plasma instabilities. For future high energy accelerators beyond energies obtainable in a single stage, the LPA design will rely on sequencing multiple stages, each driven by its own laser to supply fresh laser pulses. This coupling of LPA stages has never been demonstrated experimentally to date, and the experimental investigation of the staged LPA is the focus of this thesis.

2.6 Summary and Conclusions

This chapter provided the basic physics of LPAs. The concept of CPA laser systems, the breakthrough technology that allowed experimental investigation of LPAs, was introduced. The discussion on the production of ultraintense short laser pulses was followed by the theoretical framework of a Gaussian pulse propagation. The diffractive nature of laser pulses is a limitation to the acceleration length in the LPA if not compensated. The theory of laser guiding in a plasma channel waveguide to mitigate diffraction was discussed.

The physics of plasma wave excitation and its properties such as plasma wavelength, amplitude, and phase velocity were discussed. The properties of the plasma wave are mostly determined by laser intensity and plasma density. In the linear regime, the plasma waves are symmetric in accelerating/decelerating and focusing/defocusing fields, providing a $\lambda_p/4$ phase region for electron acceleration. In the nonlinear regime, the plasma wavelength increases and a larger phase region is suitable for electron acceleration. The concept of dephasing, which is another challenging limitation to the acceleration length in an LPA, was introduced. Then, three e-beam production methods using self-trapping, ionization of high Z atoms and negative density gradient were discussed.

Electron acceleration was then discussed in the context of energy transfer from the laser to electrons via the plasma medium. Scaling laws of electron energy gain in a single LPA module were presented. In the linear regime, the acceleration length is limited by dephasing if laser diffraction is controlled. In the nonlinear regime, dephasing and depletion lengths are similar. However, electron injection is easily triggered in the nonlinear regime, producing dark current in the LPA. The quasi-linear regime is a suitable regime for post-acceleration because the laser energy is efficiently transferred to the plasma without trapping electrons. For high energy accelerators beyond energies obtainable in a single module, the LPA design will rely on staging multiple modules to achieve desired e-beam properties and energy. This is the motivation behind the staging experiment.

Chapter 3
Staged Laser-Plasma Accelerator: Introduction

3.1 Introduction

In this chapter, the experimental design and setup of the staging experiment are presented. The central goal of the staging experiment is to drive two LPA modules with two independent laser pulses and transport the e-beam between them. This will demonstrate that a fresh laser pulse can be supplied to a sequenced module and pump depletion can be overcome. Furthermore, staged acceleration allows independent control of electron injection and acceleration. This is a first step for applications such as high energy accelerators [27, 28].

Section 3.2 discusses the experimental design and goals. Section 3.3 introduces the experimental setup. A new laser beamline was constructed for the staging experiment at the LOASIS facility. The overview of the beamline is presented first, then the LPA target system consisting of two LPA modules and a plasma mirror are presented. High power laser diagnostics and e-beam diagnostics are discussed at the end. Experimental results relevant to staging will be presented in the following chapters.

3.2 Experimental Design

A schematic of the staging experiment is shown in Fig. 3.1. Electron beams are produced in the first module (1st module), and the purpose of the second module (2nd module) is to post-accelerate the e-beam. The 1st module is driven by laser pulse 1 and the 2nd module is driven by laser pulse 2. Laser pulse 2 cannot be reflected off a conventional mirror near its focus because the intensity exceeds the damage threshold of conventional optics. A plasma mirror is employed to reflect the laser pulse near the focus and keep the distance between the modules to a minimum. The plasma mirror will be discussed in detail in Chap. 5 along with experimental results [27, 57, 58].

S. Shiraishi, *Investigation of Staged Laser-Plasma Acceleration*,
Springer Theses, DOI: 10.1007/978-3-319-08569-2_3

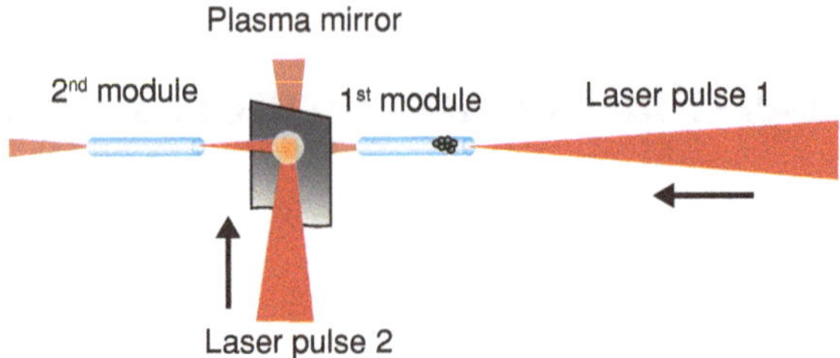

Fig. 3.1 A schematic of the staging experiment. Electron beams are produced within the 1st module, and the purpose of the 2nd module is to provide efficient acceleration. The second laser pulse is coupled in to the acceleration beamline using a plasma mirror

3.3 Experimental Configuration

The TREX laser system was introduced in Sect. 2.2 and the staging beamline is shown in Fig. 3.2.

Each pulse from the TREX laser system was split into two to drive the two modules. Pulses are transported into the staging beamline from the lower right corner in Fig. 3.2. The pulse length was adjusted by the compressor placed upstream of this figure before splitting. The polarization of the laser pulses was rotated from

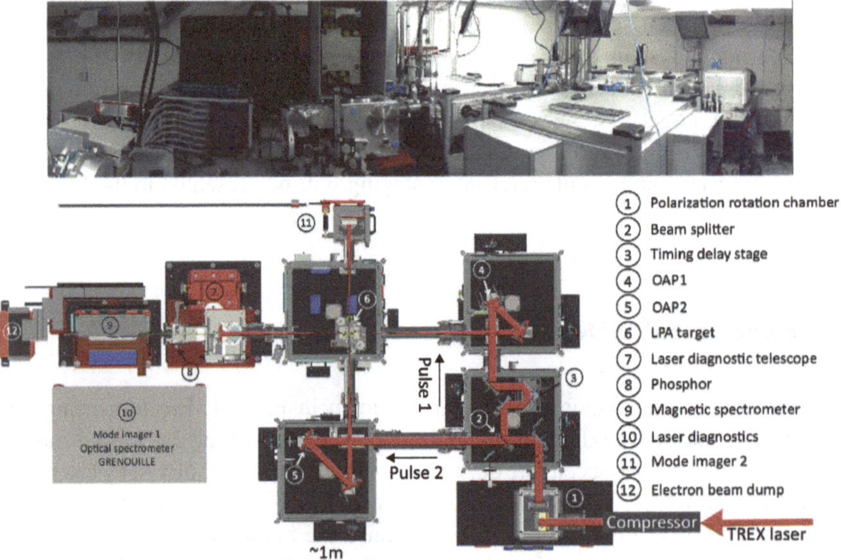

Fig. 3.2 Staging experimental setup. *Top* A photo of the laser transport chambers, laser diagnostic and magnetic spectrometers. *Bottom* CAD model of the experimental setup indicating critical components in the beamline

p-polarization to s-polarization in the chamber indicated as ①. This was to increase the laser reflectivity off the plasma mirror, which will be discussed in Chap. 5. Laser pulse energy before splitting was ~1.7 J. The beam splitter ② reflects 60 % (~25 TW at 40 fs) of the laser energy and transmits 40 % (~15 TW at 40 fs) of the energy. The reflected pulse of 1 J was laser pulse 1 and the transmitted pulse of 0.7 J was laser pulse 2. The beam splitter was made of fused silica and was 10 mm thick at 0°. Laser pulse 2 undergoes significant modulation in the beam splitter at high power, which will be discussed in Chap. 6. Laser pulse 1 is used to produce e-beams in the 1st module. Laser pulse 2 was reflected off the plasma mirror and used to excite wakefields in the 2nd module for the post-acceleration of the e-beam produced in the 1st module. The arrival timing of the two laser pulses at the target could be controlled with the delay stage indicated as ③. The delay stage could change the path length of laser pulse 1, adjusting the arrival time of laser pulse 1 with respect to laser pulse 2. Both laser pulses were focused using off-axis parabolic mirrors (OAPs) of 2 m focal lengths (④ and ⑤) to spot sizes of $r_0 \simeq 20\,\mu\text{m}$. These laser pulses from the TREX laser system were focused onto plasma targets indicated as ⑥.

A photo of the LPA target assembly is shown in Fig. 3.3. The target system (1st module, 2nd module and the tape for the plasma mirror) was placed on a hexapod that allows six-axis adjustment of position and angle. The 1st module and the 2nd module are indicated as ① and ②. Laser pulse 1 and laser pulse 2 are indicated as ③ and ④. Laser pulse 2 approached the target system from 90° and was coupled into the accelerator beamline with the plasma mirror indicated as ⑤. The tape-drive system for the plasma mirror is indicated as ⑥. A VHS tape was used as a target to form a plasma mirror. After each laser shot, the tape was spooled to provide a fresh surface at the interaction area for the following laser pulse. The distance between the 1st module and the front surface of the plasma mirror was 10 mm. The distance between the plasma mirror and the 2nd module was 13.5 mm, placing the two modules ~23.5 mm apart.

The target components were pre-aligned on the acceleration axis. The modules were placed on manual stages indicated as ⑦, which allowed position and angle adjustments with six fine-adjustment screws. After the target system was aligned in air, the chamber was pumped to vacuum. The final target alignment with respect to the laser was performed using a continuous wave diode laser and adjustments to the hexapod with micron precision in translation and ~0.1 mrad in angle. Then, the alignment between the laser and the plasma channel formed in the capillary was checked by guiding low power TREX laser pulses (<10 mJ) through the channel using techniques which will be described in Chap. 6.

Both LPA modules used capillaries of 33 mm in length and 200–250 μm in diameter, machined into sapphire plates and assembled at the LOASIS facility. The exact geometry of the capillaries will be described along with each experimental result. An example of a capillary and a gas jet (for plasma density tailoring) machined on sapphire plates is shown in Fig. 3.4. A neutral gas (H_2 or mixed gas of N_2 and He) was flowed through slots located at the ends of the capillary. A high-voltage (HV) electric discharge was applied through cables indicated as ⑧ in Fig. 3.3 to ionize the gas and form a plasma. The basic physics of plasma channel formation in the capillary will

Fig. 3.3 A photo of the staging target system. The target assembly is placed on a hexapod that allows precise alignment of the six-axes

be discussed in Sect. 6.3. More detailed theory and experimental investigations of capillary discharge waveguides are found elsewhere [47, 59, 60].

The post-interaction high power laser pulses were attenuated, focused and diagnosed. A photo and a model of the attenuation and focusing optics are shown in Fig. 3.5. The laser pulses were attenuated with an optically flat glass wedge, indicated as ②, placed 1.28 m downstream of laser pulse 1 focus. The wedge reflected up 1 % of the energy for diagnosis and the remainder was transmitted to a laser pulse dump, indicated as ⑥. This wedge can be rotated out of the beamline to let the e-beam through to the e-beam diagnostics. The reflected fraction of the laser pulse propagated through a window separating vacuum and air and was weakly focused by a Cassegrain style telescope.

The telescope consisted of a pair of concave, ③, and convex, ④, mirrors and focuses the laser pulses without introducing significant aberrations. The optics were silver coated to minimize the reflectivity dependence on laser wavelength. The concave mirror was 152.4 mm in diameter with a radius of curvature $R_c = 1200$ mm, placed at 6.50° (half-angle) with respect to the laser axis. The convex mirror was

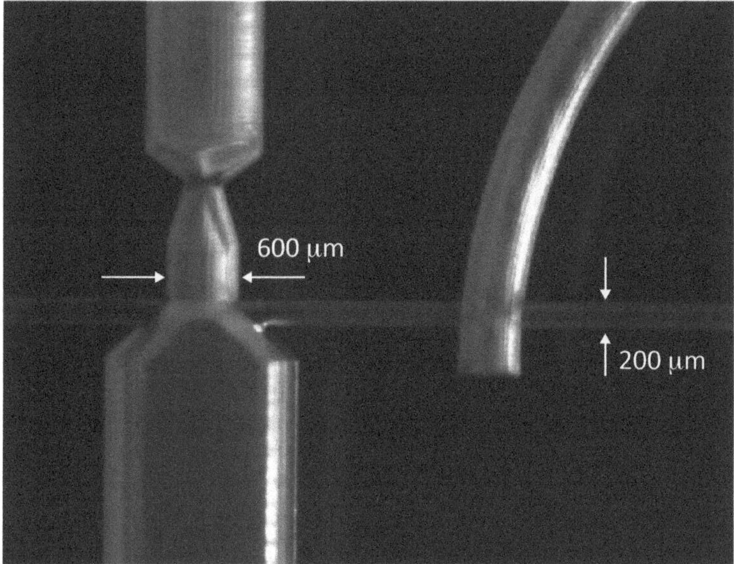

Fig. 3.4 An picture of a capillary and a gas jet machined on sapphire plates. The *horizontal line* is the capillary of 200 μm diameter. The structure on the *left* is the gas jet nozzle of 600 μm diameter. High pressure gas is supplied from the *top* of the nozzle. The *curved slot* on the *upper right* is a low pressure gas inlet

101.6 mm in diamter with $R_c = -820$ mm, placed at 12.85°. The concave mirror was placed 2 m from the target, and the convex mirror was placed 0.5 m from the concave mirror. The concave mirror determined the angular acceptance of the laser pulses and was f/13. This telescope provided a transverse magnification of 3.3. The telescope optics were placed on a linear translation stage, ⑨, allowing optical imaging from the entrance of the 1st module to the exit of the 2nd module.

The reflected pulse that was weakly focused through the telescope was again attenuated with an optically flat wedge, indicated as ⑤, for diagnosis. The energy of the laser pulse transmitted through this second wedge was measured with an energy meter which was cross-calibrated with the input laser energy. This allows measurements of laser energy transmission through the capillaries. The reflection of the second wedge was transported out of the plane for further diagnosis on the diagnostic table shown in Fig. 3.2 ⑩. The layout of the diagnostic table is shown in Fig. 3.5.

The optical image of the laser at the target was recorded using a 12-bit charge-coupled device (CCD) camera, referred to as a mode imager. To allow an imaging range of ∼15 cm at target, the laser path length in the telescope was adjusted with a combination of the translation stage shown in Fig. 3.5 and the telescope linear stage. Before the mode imager, the laser pulse was split with a pellicle and optical spectra were also measured. For some experiments, laser pulse durations were measured using GRENOUILLE [61].

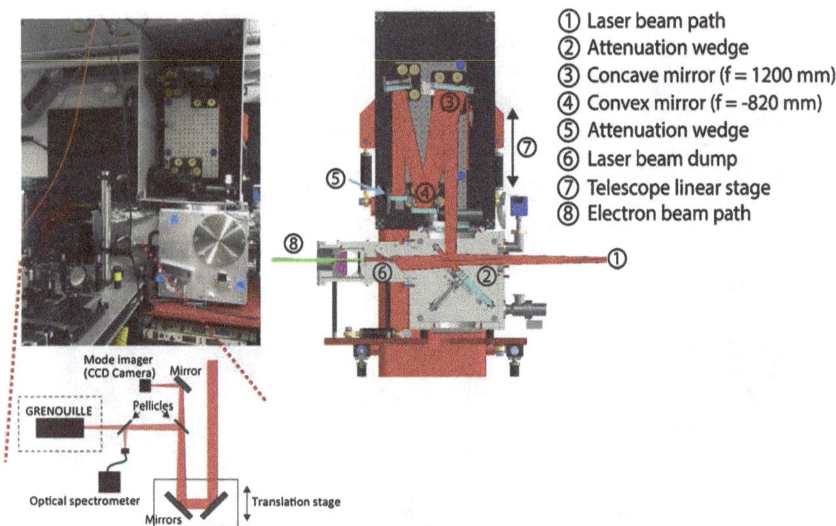

Fig. 3.5 A photo (*left*) and a CAD model (*right*) of the laser diagnostic telescope next to each other. The laser pulses reflected off the second attenuation wedge propagate out of the plane and onto the diagnostic table. The optical components on the diagnostic table are shown in *lower left*. Laser mode and spectra were measured for every shot. GRENOUILLE was used to measure pulse duration in some experiments [61]

A magnetic spectrometer to measure electron energy distribution was placed 2.36 m downstream of the laser pulse 1 focus. A 1 T dipole magnet deflected the electrons onto phosphor screens imaged by three synchronously triggered CCD cameras, enabling single-shot detection of electrons with energies in the range 60–2400 MeV [62]. This diagnostic provided charge-density images of the e-beam with energy in one direction and angle in the undispersed plane. The momentum resolution was a function of momentum and divergence. For electron momentum of $\sim$60–300 MeV/c, $<1\,\%$ momentum resolution per milliradian divergence was achieved. There was a limiting aperture of diameter 15.24 mm, 2.0 m downstream. The electron acceptance aperture was defined by this geometry and was $\pm$4 mrad.

3.4 Summary and Conclusions

The central goal of the staging experiment is to demonstrate that fresh laser pulses can be supplied to overcome pump depletion, which is a critical step towards development of LPA based high energy accelerators. The design, procurement, installation and commissioning of the staging experimental setup was a part of this thesis. In the staging setup, each TREX laser pulse was split into 25 TW (laser pulse 1) and 15 TW (laser pulse 2) pulses to drive two LPA modules independently. Laser pulse

1 was used to produce e-beams in the 1st module and laser pulse 2 was reflected off the plasma mirror and was used to excite wakefields for post-acceleration of the e-beams. The setup also allowed laser and e-beam diagnostics including laser mode, optical spectra and e-beam spectra. The staging target assembly including two capillaries and the tape-drive based plasma mirror was also discussed. The experimental results of the three critical components will be presented in the following chapters: e-beam production in the 1st module will be presented in Chap. 4, plasma mirror characterization will be presented in Chap. 5 and wake excitation in the 2nd module will be presented in Chap. 6.

Chapter 4
Injection Module

4.1 Introduction

In this chapter, experiments relevant to e-beam production in the 1st module are discussed. The staging experiment requires stable e-beams with small energy spread to observe the effect of post-acceleration in the 2nd module. Electron beam production depends on properties of plasma waves such as shape, amplitude and phase velocity. These properties are in turn functions of laser intensity, spatial and temporal profiles as well as plasma density and density profiles. While various e-beam properties have been demonstrated from LPAs, precise knowledge of laser dynamics in the plasma, plasma evolution and injection thresholds are yet to be understood.

Section 4.2 discusses experimental results of an injection study with pulses $\leq$40 TW ($\leq$1.7 J). Experiments were performed in the undulator beamline which uses pulses from the TREX system without splitting and have been operating since 2005. The laser focusing geometry (f-number) is the same and capillary geometries are the same or similar to the staging experiment. The experiments in the undulator line allowed investigation of e-beam production in an established instead of the newly installed staging beamline. Three categories of experiments are discussed. The first category employed a hydrogen filled capillary waveguide. The dominant mechanism for e-beam production is self-trapping and required high a_0. The second category used a capillary waveguide filled with a mixed gas of N_2 and H_e. This experiment produced e-beams via ionization injection which lowered the a_0 threshold to an accessible value in the staging experiment. The third category of experiments used a capillary waveguide with gas jet (jet+cap) embedded in the upstream region. This target tailored the plasma density so as to facilitate self-trapping near the jet and reduce the e-beam energy spread. The basic physics behind these injection mechanisms were introduced in Chap. 2.

Section 4.3 discusses characterization of LPA produced e-beams. Observation of a long range femtosecond structure in broad energy spread e-beams and emittance measurements based on X-ray spectra and divergence measurements from a magnetic spectrometer are discussed. For successful coupling and the post-acceleration of e-beams in staged LPAs, understanding e-beam properties produced in the 1st module are critical.

S. Shiraishi, *Investigation of Staged Laser-Plasma Acceleration*,
Springer Theses, DOI: 10.1007/978-3-319-08569-2_4

Section 4.4 discusses the initial experimental results on e-beam production in the staging setup using 25 TW pulses. Experiments were performed based on studies discussed in Sect. 4.2. Section 4.5 discusses e-beam dynamics between the modules. Implications on requirements for the e-beam properties produced in the 1st module and are compared with measured properties discussed in Sect. 4.3. In a staged LPA, production of a high quality e-beam is more critical than ever since the e-beam must be transferred to the sequential stage and accelerated in a controlled manner.

4.2 Experiments on Electron Beam Production at 40 TW

4.2.1 Experimental Configuration

In the undulator beamline, the laser energy can be as high as 1.7 J on target with 40 TW peak power since the pulses are not split. The experimental layout is shown in Fig. 4.1 and the laser and plasma channel parameters are listed in Table 4.1. TREX laser pulses were focused onto plasma targets with a focal spot size of $r_0 \sim 20\,\mu\text{m}$ using an off-axis parabolic mirror at f/20, the same as the staging experiment. The input energy of each laser pulse was measured by a photodiode that was calibrated to the energy on target. The energy of the the laser pulses transmitted through the capillary was measured by a pyroelectric energy meter that was cross-calibrated to the input photodiode.

All experiments presented in this chapter utilized capillary waveguides that were similar to those used in the staging experiment introduced in Sect. 3.3. The

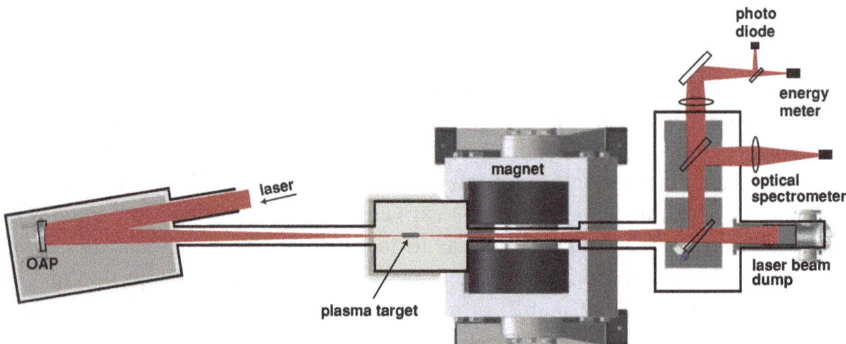

Fig. 4.1 The layout for e-beam production experiments in the undulator beamline. Pulses from TREX laser system were focused onto the plasma target with a focal spot size of $r_0 \sim 20\,\mu\text{m}$ using an off-axis parabolic mirror at f/20. Laser pulses exiting the capillary were attenuated by reflection off two optically flat wedges. Then the energy and optical spectrum of the laser pulses were measured with the diagnostics shown. The spectra of e-beams coming out of the plasma were measured using a magnetic spectrometer. Figure taken from [23]

Table 4.1 Parameters used for the e-beam production experiments in the undulator beamline

Data set	Gas species	Laser E (J)	a_0	n_0 (cm^{-3})	$\Delta\tau$ (fs)	Z_f (mm)	P/P_c	Injection type
1	H_2	1.6	1.4	3.4×10^{18}	44	4.0	3.8	Self-trapping
2	5 % N_2 + H_e	1.6	1.4	2.3×10^{18}	44	4.0	2.6	Ionization
3	5 % N_2 + H_e	1.3	1.2	2.3×10^{18}	44	4.0	2.1	Ionization
4	5 % N_2 + H_e	1.0	1.1	3.6×10^{18}	44	4.0	2.6	Ionization

Gas species, laser energy, a_0, n_0, pulse duration $\Delta\tau$ at FWHM, laser focus location with respect to the capillary entrance Z_f, and P/P_c for self-focusing threshold are listed

exact geometry of the targets differ for different injection mechanisms and are described in the corresponding sections. The peak discharge current was ~200–250 A and ~400 ns wide at FWHM. The basic physics of plasma channel formation in the capillary will be discussed in Sect. 6.3. More detailed theory and experimental investigations of capillary discharge waveguides are found in [47, 59, 60].

The energy of the e-beam produced by the LPA was measured by a magnetic electron spectrometer [63]. A 1.2 T dipole magnet deflected the electrons onto phosphor screens imaged by four synchronously triggered CCD cameras, enabling single-shot detection of electrons with energies in the range 0.01–0.14 GeV and 0.17–1.1 GeV. Electron beam charge was obtained from the phosphor screens which were cross-calibrated against an integrating current transformer. This diagnostic provided charge-density images of the e-beam with energy in one direction and angle in the undispersed plane.

4.2.2 Electron Beam Production via Self-trapping

When the laser a_0 is large and the amplitudes of the excited wakefield approaches the wavebreaking limit, e-beams can be produced through the self-trapping of background plasma electrons as discussed in Sect. 2.4.3. Hydrogen gas was employed to form a plasma channel in a 250 μm diameter 33 mm long capillary. Laser and plasma parameters were 1.6 J, focal spot size $r_0 \simeq 20\,\mu\text{m}$, Strehl ratio 0.9, $a_0 \simeq 1.4$, plasma density $n_0 \simeq 3.4 \times 10^{18}\text{cm}^{-3}$, channel matched spot size $r_m \simeq 34\,\mu\text{m}$, ratio of power normalized by the critical power for self-focusing $P/P_c \simeq 3.5$, and pulses arrived ~130 ns after the peak of discharge current. Other experimental parameters for this study are summarized as Data set 1 in Table 4.1.

An example of the e-beam spectra is shown in Fig. 4.2. For this e-beam, the charge was $Q \simeq 13\,\text{pC}$, mean energy of $\bar{E} \simeq 269\,\text{MeV}$ and divergence was $\theta_{\text{FWHM}} \simeq 1.0\,\text{mrad}$. The energy of the charge density peak was $E_{\text{peak}} \simeq 256\,\text{MeV}$ with energy spread of $\sigma_{\Delta E/E} \simeq 30\,\%$. Using this injection method dominated by self-trapping, on average, $Q \simeq 7 \pm 4\,\text{pC}$, $\bar{E} \simeq 243 \pm 53\,\text{MeV}$, $\theta_{\text{FWHM}} = 1.7 \pm 0.2$ ($\sigma_\theta = 0.8 \pm 0.03$) mrad, $E_{\text{peak}} \simeq 236 \pm 54\,\text{MeV}$, and $\sigma_{\Delta E/E} \simeq 30\,\%$. The large

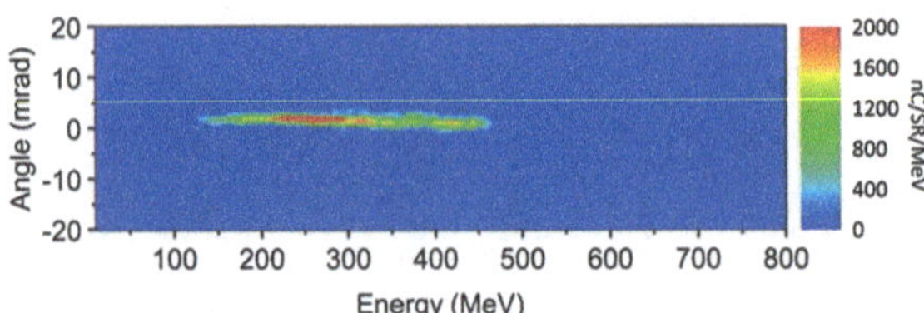

Fig. 4.2 An example e-beam spectrum produced using $a_0 \simeq 1.4$ and $n_0 \sim 3.4 \times 10^{18}\,\text{cm}^{-3}$. The dominant injection mechanism was self-trapping. The e-beam charge was 13 pC, mean energy 269 MeV and divergence 1 mrad at FWHM

$\Delta E/E$ suggests that electron trapping is not terminated through reduction of wake amplitude, or the acceleration length is not optimized to the dephasing length. The e-beams were not observed with this configuration when $n_0 < 2.6 \times 10^{18}\,\text{cm}^{-3}$ or pulse energy $\leq$1.3 J on target. Since injection must be achieved with ~1 J in the staging experiment, self-trapping injection was not feasible.

4.2.3 Electron Beam Production via Ionization of N_2

Electron beam injection was observed within a wider range of laser intensity and plasma density using a mixture of 5 % N_2 balanced with H_e gas in the capillary, attributed to ionization injection as introduced in Sect. 2.4.3. It was discussed that the ionization threshold of inner bound electrons in nitrogen atoms is $a_0 \sim 1.5$–2.0. For pulses from the TREX system, a_0 is less than or equal to 1.5 even at full laser power. Therefore, the injection was an interplay of input a_0 and relativistic self-focusing which increased a_0 in the plasma. The strength of self-focusing is estimated with P/P_c ($P \gg P_c$ indicates greater self-focusing effect) as discussed in Sect. 2.3.3. Experiments presented in this subsection were performed using the same r_0 and $\Delta\tau$ as the self-trapping experiment while varying the laser energy and plasma density to change a_0, n_0, r_m and P/P_c. Figure 4.3 shows P/P_c for three different laser powers used in the experiment. For a lower laser power, the plasma density was raised to increase P/P_c and achieve stronger self-focusing in the plasma. By adjusting P/P_c, e-beam production has to be realized at <25 TW (blue dot-dashed line in Fig. 4.3) to prepare for the staging experiment.

To confirm the lower threshold for ionization injection than that of self-trapping, the plasma density was reduced while maintaining a_0. The experimental parameters are listed as Data set 2 in Table 4.1. With the same laser intensity as Data set 1 ($a_0 \sim 1.4$), e-beams were produced at $n_0 \sim 2.3 \times 10^{18}\,\text{cm}^{-3}$, below the observed threshold using H_2 gas. An example e-beam spectrum is shown in Fig. 4.4a. Injection was observed in 100 % of laser shots and low energy, large divergence electrons were produced along with a higher energy component with a narrower divergence. Averaged e-beam parameters are shown in Fig. 4.5. The total charge was

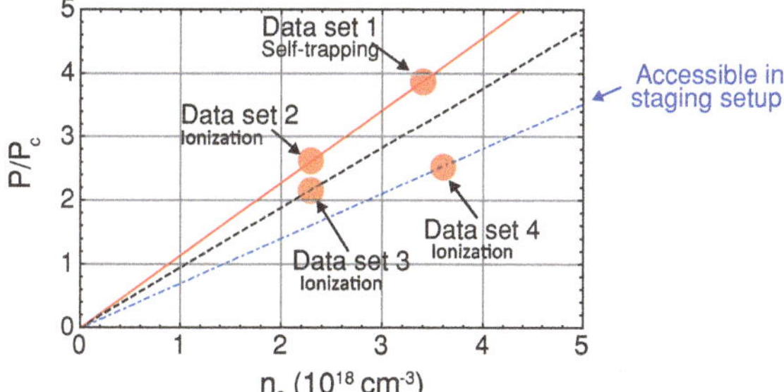

Fig. 4.3 Strength of self-focusing effect, P/P_c, as a function of plasma density n_0 for three different laser powers used in Data set 1–4. For a lower laser power, n_0 was raised to increase P/P_c and achieve stronger self-focusing in the plasma

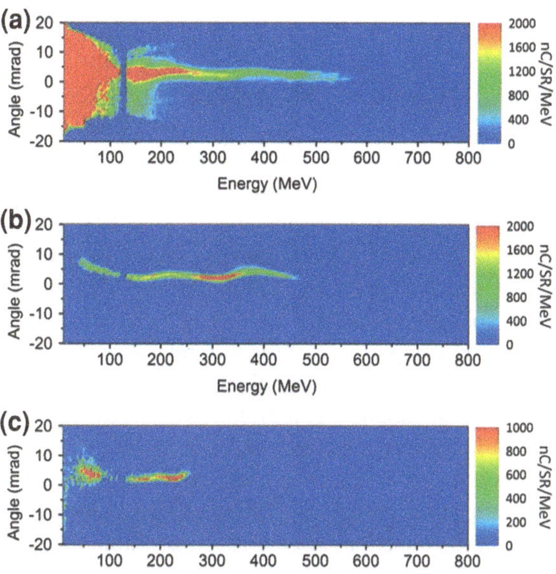

Fig. 4.4 Example spectra of e-beams produced via ionization for three different laser energies and plasma densities. The parameters were **a** $a_0 \simeq 1.4$ and $n_0 \sim 2.3 \times 10^{18}\,\text{cm}^{-3}$ **b** $a_0 \simeq 1.2$ and $n_0 \sim 2.3 \times 10^{18}\,\text{cm}^{-3}$ **c** $a_0 \simeq 1.1$ and $n_0 \sim 3.6 \times 10^{18}\,\text{cm}^{-3}$

$\sim$86 $\pm$ 45 pC, significantly greater than the charge of e-beams produced by self-trapping in H_2 gas (7 $\pm$ 4 pC). Due to the large charge in the low energy component, $\bar{E} \simeq 144 \pm 59$ MeV even though the maximum energy was 602 $\pm$ 99 MeV.

When the input laser intensity was reduced to $a_0 \sim 1.2$ (1.3 J) while maintaining the same plasma density (Data set 3 in Table 4.1), electron injection was at threshold. An example e-beam spectrum is shown in Fig. 4.4b, and the average parameters are shown in Fig. 4.5. The total charge was $\sim$12 $\pm$ 6 pC, the average energy $\bar{E} \simeq 266 \pm$ 76 MeV and the maximum energy was 470 $\pm$ 9 MeV. The e-beam was observed only in $\sim$50 % of laser shots. However, e-beams were narrower divergence and contained less low energy electrons, more desirable than those obtained using $a_0 \sim 1.4$.

When the input laser intensity was further reduced to $a_0 \sim 1.1$, equivalent laser energy (1 J) to the staging experiment, injection required higher n_0 to enhance self-focusing. With the same plasma density as Data set 2 and 3 ($n_0 \sim 2.3 \times$

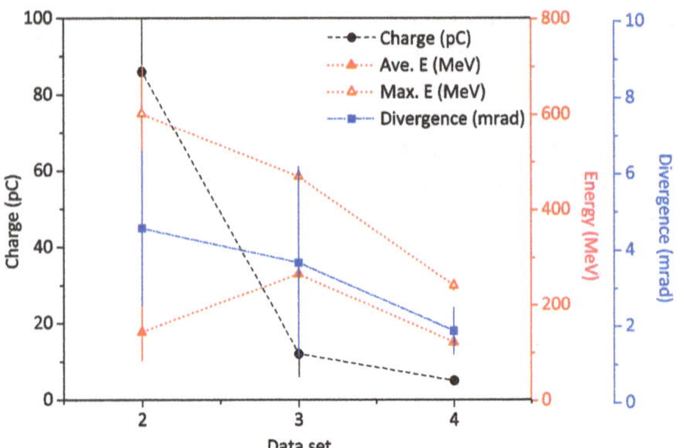

Fig. 4.5 Averaged e-beam parameters observed for Data sets 2–4 listed in Table 4.1. The lines connecting the symbols are shown only for clarity

10^{18} cm^{-3}), no e-beam was produced, attributed to the lower self-focusing ($P/P_c \simeq 1.6$). Electron beams were produced when n_0 was raised to 3.6×10^{18} cm^{-3} ($P/P_c \geq 2.6$) as listed as Data set 4 in Table 4.1. An example e-beam spectrum is shown in Fig. 4.4c and e-beam parameters are shown in Fig. 4.5. The total charge was $\sim 5 \pm 1$ pC, the average energy $\bar{E} \simeq 123 \pm 17$ MeV and the maximum energy was 242 ± 7 MeV. Injection was observed in ~ 66 % of shots within $n_0 \sim 3.6$–4.9×10^{18}cm^{-3}, suggesting that ionization injection is at threshold with the staging experimental parameters.

These examples with three different laser energies illustrate that as a_0 decreases, e-beam production requires higher n_0 to self-focus the laser pulse and achieve higher a_0 in plasma than in vacuum. In experiments, plasma density will be limited by the amount of gas that can be flowed through the capillary module while maintaining pumping in the vacuum chamber and reliably discharging the gas across the capillary. Operating near the injection threshold produces e-beams with a narrower divergence but reduces injection probability. In addition, the charge and the maximum electron energy also decrease with decreasing a_0. Electron beam production in the staging experiment will be similar to Data set 4 in Table 4.1. The expected e-beams are of maximum energy ~240 MeV with large energy spread with ~5 pC, and ionization injection will be at threshold. In the next subsection, tailoring of plasma density profile to facilitate self-trapping injection to stabilize e-beam properties is discussed. At the end of this chapter, experiments to facilitate ionization injection using the tailored plasma density profile in the staging setup are presented.

4.2.4 Electron Beam Production with Tailored Plasma Density

In this section, experiments on reducing e-beam energy spread and energy fluctuation, and increasing the injection probability using a negative plasma density gradient are presented. This experimental results were published in Gonsalves et al. [23]. Electron injection is facilitated by increasing a_0 through self-focusing and intentionally decreasing the phase velocity of the plasma wave through a negative plasma density gradient as introduced in Sect. 2.4.3. Helium gas was used for the

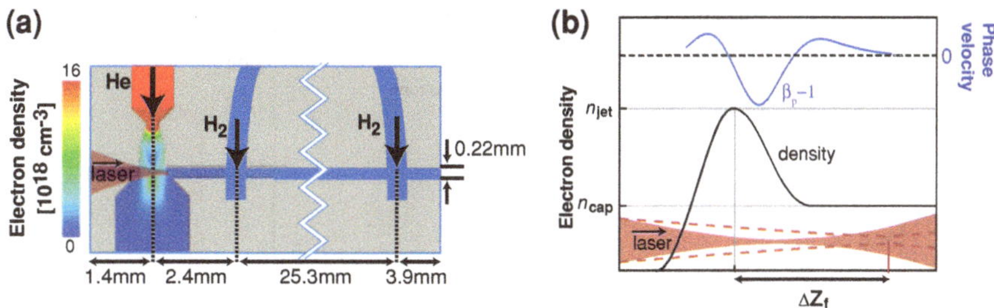

Fig. 4.6 **a** The target schematic representation and gas profile calculated by a commercial fluid-dynamics code, where a supersonic gas jet has been embedded into a capillary that is filled with hydrogen gas. **b** Illustration of the laser focusing onto the plasma target and of the basic physics: the longitudinal on-axis density profile in the region of the gas (*black*), the location of the vacuum focus with respect to the center of the gas jet ΔZ_f (*red dashed line*), the self-focused laser focal position (*red shaded area*) and the wake phase velocity (*blue line*). Figures taken from [23]

gas jet and hydrogen gas was used for the plasma channel. The laser parameters were $r_0 = 22\,\mu\text{m}$, $\Delta\tau \simeq 38\,\text{fs}$, $a_0 \simeq 1.5$ and plasma density was a function of distance.

The longitudinal density was tailored using a supersonic gas jet embedded near the entrance of the capillary [23]. The schematic is shown in Fig. 4.6a. The result was a high density region (gas jet) of $550\,\mu\text{m}$ in length connected to a lower density region (capillary) as shown in Fig. 4.6b. The total length was 33 mm. The peak density of the gas jet region was $n_{\text{jet}} \sim 7 \times 10^{18}\text{cm}^{-3}$ and the on-axis density of the plasma channel was $n_{\text{cap}} = n_0 \sim 1.8 \times 10^{18}\text{cm}^{-3}$. The high density n_{jet} region self-focuses the laser pulse and excites a large amplitude wakefield. The lower density n_0 region is below the threshold of injection and accelerates e-beams to roughly the dephasing length, L_d. The phase velocity of the wake as a function of the propagation distance for this density profile is shown in Fig. 4.6b. In this experiment, the high density region performed two critical functions: focusing the drive laser via relativistic self-focusing and slowing the phase velocity of the plasma wave. Simulations using INF&RNO framework [36] showed an increase of a_0 to approximately 3.5 although the input laser pulse was $a_0 \simeq 1.5$ in vacuum as shown in Fig. 4.7 (black dash-dot line). The wake phase velocity (red dashed line) is reduced in the region where the density is decreasing and the laser intensity is increasing, and trapping of particles is observed in the grey shaded region. The contribution to the phase-velocity reduction by self-focusing can be determined by comparing the wake phase velocity for vacuum $a_0 \approx 1.5$ (red dashed line) with a vacuum $a_0 \approx 0.1$ (red solid line), because for the latter self-focusing can be ignored. It can be seen that self-focusing causes a significant reduction in wake phase velocity by exciting a large bubble and extending the λ_{NP}. The gradient at the end of high density plasma region also increased $\text{d}\lambda_{\text{NP}}/\text{d}z$, hence facilitating injection by both increasing the intensity and by slowing down the plasma wave phase velocity. After the focus, the laser defocuses and reduces a_0. Simultaneously, the phase velocity of the wake increases, terminating the injection. This termination of the injection leads to electron injection only near the

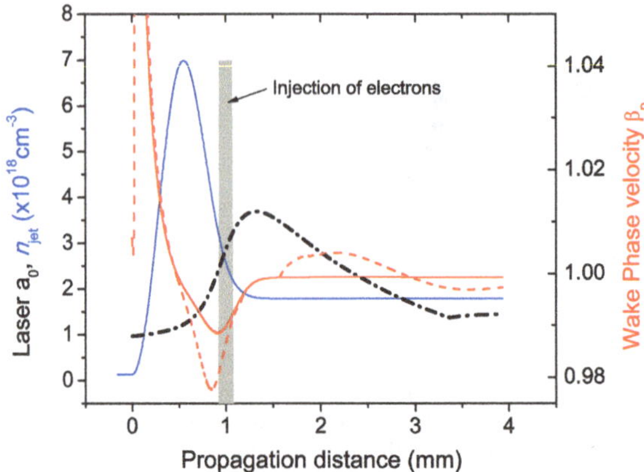

Fig. 4.7 Simulated evolution of a_0 (*black dash-dot line*) as a function of propagation distance for $\Delta Z_f = 1.6$ mm and $n_{jet} = 7 \times 10^{18}\,\mathrm{cm}^{-3}$. The blue line is the longitudinal density profile. The phase velocity reduction for $a_0 \approx 1.5$ (*red dashed line*) was greater than that for $a_0 \approx 0.1$ (*red solid line*). This was due to self-focusing increasing the nonlinear plasma wavelength. The electron trapping occurred in the shaded grey area. Figure taken from [23]

high plasma density region. Since post-acceleration lengths for these electrons are the same, the resulting e-beam has a narrower energy spread than e-beam produced via self-trapping or ionization injection without the gas jet.

Injection using a negative density gradient produced an e-beam with a much narrower energy spread and with 100 % injection probability, which was not observed for self-trapping or ionization injection in the previous subsections. Measured e-beam spectra are shown in Fig. 4.8a. The e-beam energy and charge were tuned by adjusting the laser focus position with respect to the jet center ΔZ_f and the gas jet plasma density n_{jet}. The self-focusing is maximized for $\Delta Z_f \approx 1.6$ mm and decreases on either side. The higher n_{jet} also causes increased self-focusing. When the amount of self-focusing effect is increased, the focused laser spot size is reduced and this makes diffraction of the laser pulse in the non-matched plasma channel more severe. This reduced the average laser intensity over the channel length and reduced the final energy gain. For $\Delta Z_f = 1.62$ mm and $n_{jet} \simeq 7 \times 10^{18}\mathrm{cm}^{-3}$, the peak energy was 341 MeV with $\sigma_E \simeq 1.9$ % fluctuation, energy spread was 11 % at FWHM ($\sigma \simeq 5$ %) and divergence was 2.5 mrad at FWHM ($\sigma_\theta \simeq 1$ mrad). When the laser was focused at $\Delta Z_f = 0.65$ mm, the gas jet did not provide enough self-focusing for injection. After $\Delta Z_f = 1.62$ mm, with increasing ΔZ_f, the self-focusing decreases and the e-beam energy increases. Similarly, for lower n_{jet}, the self-focusing decreases and e-beam energy increases as seen in Fig. 4.8b. However, the e-beam charge is greater when the self-focusing is strong and injection is facilitated. Using this tailored plasma density, a small e-beam energy spread and stable energy were demonstrated with 100 % injection probability.

The stable energy and the small energy spread are particularly suitable for the staging experiment. The LPA target geometry (gas jet geometry and densities) will require modification to compensate for the lower laser power in the staging exper-

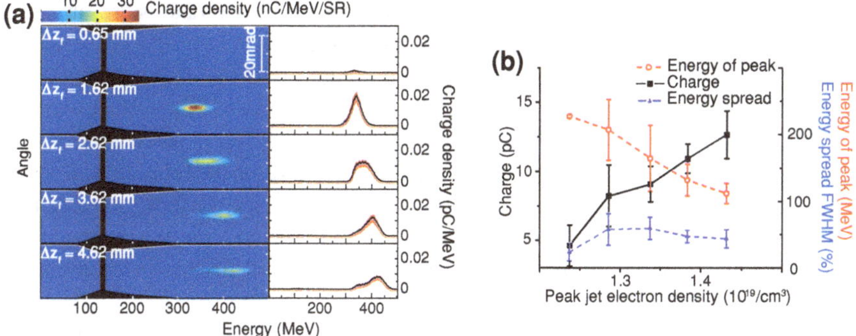

Fig. 4.8 Electron beam data from the density-tailored plasma target, demonstrating the post-injection acceleration in the capillary structure and the control over electron beam properties afforded by the density-tailored approach. **a** Averaged magnetic spectrometer images from 20 consecutive shots. The black shaded areas in each image represent the regions not covered by the spectrometer cameras. Lineouts of the mean (*black curve*) and the standard deviation (*red area*) are on the right of each image. A Helium gas jet with $n_{\text{jet}} \approx 7 \times 10^{18}\text{cm}^{-3}$ and length (FWHM) 0.55 mm was coupled to a capillary with density $n_{\text{cap}} \approx 1.8 \times 10^{18}\text{cm}^{-3}$ for various focal locations Δz_{f}. **b** The charge (*squares*), energy (*circles*), and energy spread (*triangles*), as a function of peak jet density for $\Delta z_{\text{f}} = 0.62$ mm at capillary density $n_{\text{cap}} \approx 1.2 \times 10^{18}\text{cm}^{-3}$. The error bars correspond to the standard deviation of the data. Figures taken from [23]

iment which would require a greater self-focusing effect for injection. In the next section, measurements of a slice energy spread and emittance of self-trapped e-beams are discussed. The energy spread and emittance were studied by looking at coherent optical transition radiation and X-ray spectra from betatron emission, respectively.

4.3 Characterizations of Electron Beams

4.3.1 Slice Energy Spread Measurement

Simulations indicate that the measured energy spread of e-beams in the LPA experiments are dominated by a correlated spread, with the slice spread significantly lower. The correlated energy spread can be removed by phasing the e-beam in the correct phase of the accelerating field while the slice energy spread is an intrinsic energy spread [64]. In future staging LPAs, a slice energy spread may dominate e-beam dynamics. In this section, a measurement on slice energy spread using optical transition radiation (OTR) is presented. The results were published in Lin et al. [64].

The LPA e-beam typically resides within the first plasma oscillation and overlaps with a portion of the drive laser, resulting in momentum and density modulations [64, 65]. Measurements of the micro-structure of e-beams are difficult to observe at femtosecond resolution and requires an optical-based diagnostic. The presence and persistence of laser-induced optical scale structures on LPA e-beams are observed through coherent optical transition radiation (COTR). OTR is produced when the e-beam passes through a boundary of two materials with different dielectric properties. The e-beam passing through the boundary generates a surface current, which then emits electro-magnetic radiation which is referred to as OTR. The total energy loss of the e-beam because of this radiation depends on the relativistic Lorentz factor, γ. The radiation is directed at an angle of $\sim 1/\gamma$ relative to the direction of specular reflection. When the laser induced e-beam momentum modulation becomes charge density modulation after propagation in a vacuum, the modulation can be observed through OTR.

The OTR can be coherently or incoherently added from each electron in the e-beam. If the wavelength, λ_{OTR}, of the emitted OTR is longer than the longitudinal e-beam duration, σ_z, the radiation adds coherently. If $\lambda_{\mathrm{OTR}} < \sigma_z$, the radiation adds incoherently. The incoherent contribution to the intensity of OTR per unit wavelength is $\mathrm{d}I_{\mathrm{ITR}}/\mathrm{d}k = N \int \mathrm{d}^2 r' \mathrm{d}z' \rho(\mathbf{r'}, z) \, |E(\mathbf{r} - \mathbf{r'})|^2$ where ρ is the normalized charge distribution, N the number of electrons, and $\mathbf{r}$ and z the transverse and longitudinal coordinates [38,64,66]. The coherent contribution to the radiation intensity is, $\mathrm{d}I_{\mathrm{CTR}}/\mathrm{d}k = N^2 \int \left|\mathrm{d}^2 r' \mathrm{d}z' e^{ikz} \rho(\mathbf{r'}, z) \, E(\mathbf{r} - \mathbf{r'})\right|^2$. The intensity of the radiation is, $I_{\mathrm{CTR}} \sim |E_{\mathrm{CTR}}|^2 \sim N^2$ for the coherent OTR and $I_{\mathrm{ITR}} \sim |E_{\mathrm{ITR}}|^2 \sim N$ for the incoherent OTR. The intensity of OTR is significantly greater when $\lambda_{\mathrm{OTR}} > \sigma_z$ and the radiation coherently adds. In order to observe COTR at optical wavelengths, significant transverse charge gradients and longitudinal charge structure of a scale at or below optical wavelengths ($\sim$400–900 nm) are required.

The experiment was performed on the undulator beamline and the setup is shown in Fig. 4.9a. Laser pulses with an on-target energy of 1.3 J and temporal duration of 41 fs at FWHM were used. The pulses were focused onto a supersonic gas jet or a

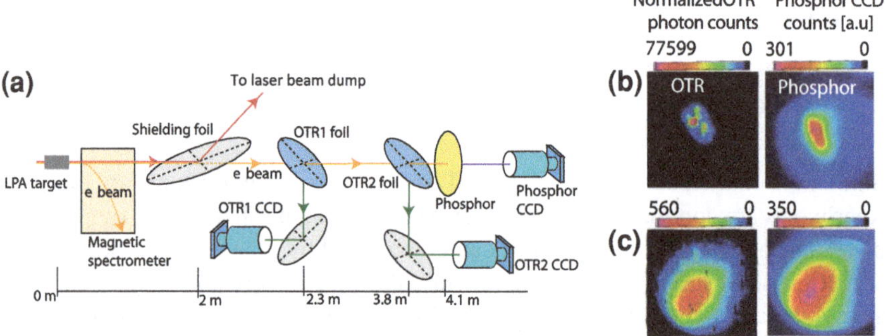

Fig. 4.9 **a** Experimental setup used to measure coherent enhancement of optical transition radiation (OTR). The surface of the OTR foils were imaged with 16-bit CCD cameras. The e-beam charge profile was measured with the phosphor screen at the end of the beamline. **b** and **c** OTR2 images and phosphor images with local coherent enhancement of $\times 1,000$ and without measurable coherent enhancement. The observed coherent enhancement at 3.8 m from the plasma target implied $\Delta\gamma/\gamma \lesssim$ 0.5 %. Figures taken from [64]

capillary with a focal spot size of $r_0 = 23\,\mu$m and Strehl ratio of $\simeq 0.85$, yielding $a_0 \simeq 1.2$. The laser ionized the gas, and excited a plasma wave in which plasma electrons were self-trapped and accelerated. The e-beam spectrum was measured by the magnetic spectrometer. The OTR measurements were subsequently taken with the electromagnet off. The laser exiting the plasma target and the e-beam were separated by $6\,\mu$m thick Mylar and $14\,\mu$m thick polycarbonate foils at $11°$. Two $5\,\mu$m thick aluminium coated Mylar foils were placed at 2.3 m (OTR1 foil) and 3.8 m (OTR2 foil) downstream from the plasma target tilted by 45° for OTR measurements. The first surface OTR emissions from both foils were recorded by 16-bit CCD cameras with $200\,\mu$m spatial resolution. Both cameras imaged the foil surfaces with a spectral sensitivity range of ~400–900 nm. At the end of the beamline, a calibrated phosphor screen recorded the e-beam position and distribution [67]. The laser and e-beam were overlapped by operating the LPA in a high density ($\sim 10^{19}$cm^{-3}) regime to enhance the e-beam modulation from the laser although this condition resulted in a broad integrated electron energy spread. The operation in the high density regime is similar to the staging experiment as will be discussed in Sect. 4.4.

The coherent enhancement of the OTR emission was defined as the ratio between the measured OTR CCD counts (integrated over the full image) to the calculated incoherent counts. The expected OTR counts in the absence of coherence were calculated based on the single-electron expression for incoherent OTR [66], the measured spatial charge profile, the averaged e-beam energy reference, and the calibrations and optical layout of the imaging system. Figure. 4.9b, c shows the measured OTR2 and phosphor images. For Fig. 4.9c, the enhancement falls to near unity where the OTR images are not only weak in counts but also match the transverse charge profile on the phosphor screen. For Fig. 4.9b, the opposite is true: the integrated OTR counts are over two orders of magnitudes higher (with local enhancement up to $\times 1,000$), and the spatial structure no longer resembles the corresponding transverse charge profile measured by the phosphor screen. The modeled coherence enhancement for a range of electron energies and slice energy spreads were compared with this experimental data. An upper limit on the slice energy spread ($\Delta\gamma/\gamma = \sigma_{\beta_z}\gamma^2$) was estimated using conservative assumptions that the charge contributing to the coherent enhancement is ~ 5 pC, and bunch duration of $2\,\mu$m ($\lambda_p/2$), and a divergence of $40\,\mu$m rad based on beam size at phosphor. For electrons contributing to the coherence, the slice energy spread was $\Delta\gamma/\gamma \lesssim 0.5\,\%$. Considering $\lesssim 10\,\%$ of the measured e-beam energy distribution in the experiment extends beyond 125 MeV, the 5 pC of charge used for the calculation represents a conservative upper limit. It is likely that there is less charge in the high-energy tail, a longer bunch at the 3.8 m foil, a larger divergence of the coherence contributing electrons, and/or a non-optimized modulation strength for density modulations, all of which reduce the modeled coherent enhancement and hence yield an even smaller slice energy spread. This measurement shows that e-beams with a broad integrated energy spread produced from an LPA can have a very small slice energy spread. If the correlated energy spread were removed by phasing the e-beams in the appropriate accelerating phase, LPA e-beams could achieve small energy spread.

4.3.2 Emittance Measurement

The emittance of the e-beam is a measure of how tightly the e-beam can be focused. For applications such as a high energy collider or a synchrotron light source, a low emittance beam is necessary to achieve high luminosity and high brightness. In this subsection, the emittance measurement is presented, and the results were published in Plateau et al. [68]. For conventional accelerators, quadrupole scan [69], optical diffraction/transition radiation interference contrast (ODTRI) [70], and the pepper-pot technique [71] are examples of emittance diagnostics. Recently, emittance has been measured on e-beams produced with LPAs for $\geq$150 MeV [72,73]. These pepper-pot measurements have limited resolution of $\sim$1 mm-mrad due to the co-propagating intense laser pulse which requires the instruments to be placed far from the electron source. Hence, these measurements are difficult to apply to low-divergence, high energy e-beams. An alternative method is to use the betatron radiation emitted from the e-beam to measure the transverse e-beam size [68, 74]. In combination with the e-beam divergence measurement, normalized emittance can be estimated.

The experimental setup is shown in Fig. 4.10a. Electron beams were produced by self-trapping of background plasma electrons in a capillary waveguide which was 250 μm diameter and 33 mm long as in the staging experiment. The laser energy on target was 1.3 J, $r_0 = 24\,\mu$m, $\Delta\tau \simeq 56$ fs, $a_0 \simeq 1.0$, and $n_0 \simeq 4$–10×10^{18}cm^{-3} using H_2 gas. The far-field, on-axis radiation between 2–20 keV was measured using a back-illuminated Si CCD camera (Andor DY420-BR-DD). The CCD camera was located 4.7 m from the LPA exit. The laser light exiting the plasma target was separated from the X-ray with the foils.

The shape of the X-ray betatron spectrum is determined by the betatron strength parameter, $\alpha_\beta \propto \sqrt{\gamma n_0} r_\beta$, where r_β is the amplitude of the betatron orbit of an electron. For a beam with radius σ_x and a given momentum distribution, the spectrum is an incoherent sum over the spectra of the individual electrons. As an electron experiences the strong plasma focusing force, it undergoes betatron oscillation and

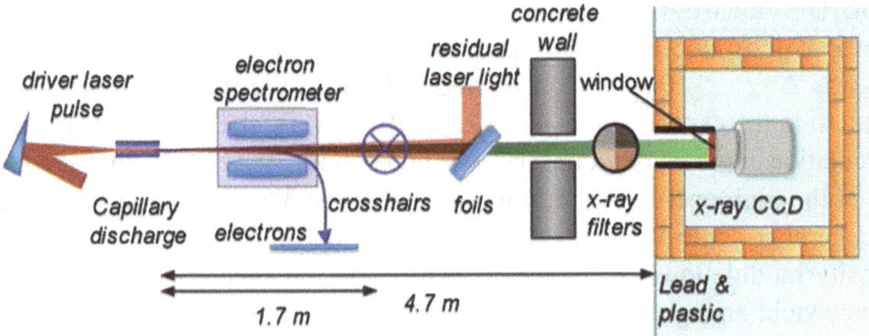

Fig. 4.10 Experimental setup for the X-ray measurement. The TREX laser was focused on a capillary target. Electron beams were measured with the spectrometer. The laser was removed using the laser pulse dump foils. The X-ray CCD camera was placed behind the concrete wall. Figure taken from [68]

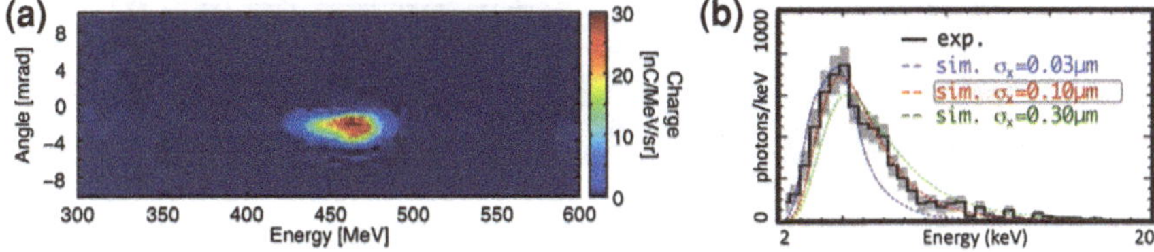

Fig. 4.11 **a** An example of measured e-beam spectra. **b** Measured X-ray spectra fitted to theoretical spectra to estimate the transverse e-beam size, σ_x for the e-beam shown in (**a**). The $\sigma_x = 0.1\,\mu$m agrees the best with χ^2/ndf$= 0.8$. Figures from [68]

emits synchrotron radiation at the wavelength of $\lambda_s = \lambda_\beta/2\gamma^2$ as discussed in Sect. 2.4. For LPAs in the bubble regime, $\lambda_\beta = (2\gamma)^{1/2}\lambda_\mathrm{p}$, and $\lambda_\beta = 0.7$ mm and $\lambda_s = 0.3$ nm (X-ray) at $\gamma \sim 1000$ [75]. Different electrons have different α_β, which blends the harmonics into a continuum whose shape is determined by σ_x and γ. Since the energy of e-beams, γ, were known from the magnetic spectrometer measurements for each shot, measured X-ray spectra were fitted to theoretical spectra for various σ_x. Figure. 4.11 shows a measured single shot X-ray spectrum from an e-beam of 463 MeV ($\gamma \sim 900$) with 2.8 % root-mean-square (rms) energy spread, 1.2 mrad rms divergence and 0.4 pC charge. The plasma density in the capillary was $n_0 = 5 \pm 2 \times 10^{18}\mathrm{cm}^{-3}$. Good agreement between the experiment and the angle integrated theoretical spectra, convolved with camera and filter response, is obtained for $\sigma_x = 0.1\,\mu$m, where χ^2/ndf statistic is 0.8. For $\sigma_x = 0.03\,\mu$m and $\sigma_x = 0.3\,\mu$m can be excluded at the 98 % confidence level. Based on the fit, the transverse e-beam size is estimated to be $\sigma_x \sim 0.1\,\mu$m [68]. Using the relation that the normalized emittance in the x-direction is $\epsilon_{N\,x} \approx \gamma\sigma_x\sigma_\theta$, it was estimated to be $\epsilon_N \sim 0.1$ mm-mrad for the e-beam shown in Fig. 4.11.

This measurement showed that the emittance of an LPA produced e-beam via self-trapping can be as low as 0.1 mm-mrad. The trajectory and hence emittance of e-beams will depend on the plasma wave amplitude and therefore on parameters including a_0 and n_0. The observed variation of $\epsilon_{N\,x} \approx 0.1$–1 mm-mrad in the data indicates that the control of the emittance through laser-plasma parameters can be explored in future experiments.

4.4 Experiments on Electron Production at 25 TW

Based on the injection study presented in Sect. 4.2, e-beam production in the 1st module was investigated in the staging setup. The requirement for the staging experiment is such that energy fluctuation is significantly less than the expected energy gain from the 2nd module and injection probability must be high enough that these e-beam fluctuations can be statistically separated from effects in the 2nd module.

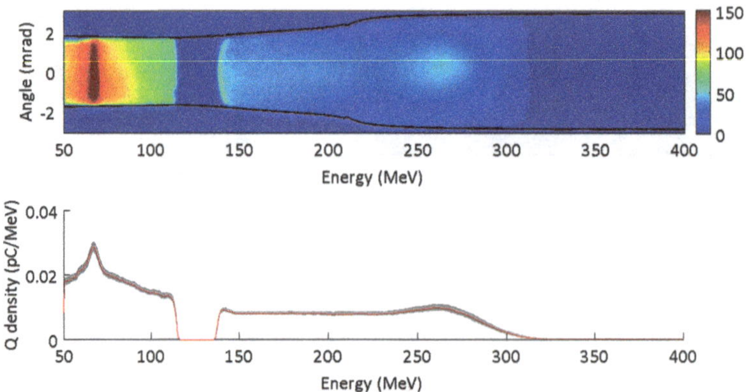

Fig. 4.12 Averaged spectrum of 69 consecutive e-beams produced via ionization injection in a capillary waveguide in the 2nd module. The charge of the e-beam was $Q \simeq 2.9 \pm 0.8$ pC, $\bar{E} \simeq 134 \pm 18$ MeV, $\theta_{\text{FWHM}} = 3.0 \pm 0.4$ ($\sigma_\theta = 1.07 \pm 0.04$) mrad, $E_{\text{peak}} \simeq 74 \pm 36$ MeV, and the rms energy spread $\sigma_{\Delta E/E} \simeq 59$ %. The rms fluctuation of E_{peak} was 53 %

From the study presented in Sect. 4.2.2, self-trapping in H_2 plasma was deemed not feasible since it required ~40 TW for our focal geometry and setup. The capillary was filled with a mixed gas of 1 % N_2 balanced with H_e to achieve ionization injection. The 1 J laser pulses were focused to $r_0 \simeq 21$ μm with Strehl ratio 0.8, $\Delta\tau \simeq 67$ fs at FWHM, $a_0 \simeq 1.1$, $n_0 \simeq 4.6 \times 10^{18}$cm^{-3}, $r_{\text{m}} \simeq 31$ μm and $P/P_c \simeq 2.7$. The laser was focused $Z_{\text{f}} = 4$ mm into the capillary waveguide and arrived at the peak of the discharge current which was ~200 A and ~400 ns wide at FWHM.

Ionization injection in a plasma channel at high n_0 resulted in a high injection probability, but with a broad e-beam energy spread and fluctuating energy. An averaged e-beam spectrum of 69 consecutive shots is shown in Fig. 4.12. On average, $Q \simeq 2.9 \pm 0.8$ pC, $\bar{E} \simeq 134 \pm 18$ MeV, $\theta_{\text{FWHM}} = 3.0 \pm 0.4$ ($\sigma_\theta = 1.07 \pm 0.04$) mrad, $E_{\text{peak}} \simeq 74 \pm 36$ MeV, and the rms energy spread was $\sigma_{\Delta E/E} \simeq 59$ %. Although the the injection probability was 100 %, the E_{peak} also fluctuated by 53 %. This fluctuation in the energy and the large $\Delta E/E$ would imply that even if these e-beams were transported to the 2nd module, change in e-beam energy from the 2nd module would not be distinguished from the injection fluctuation in the 1st module.

To improve the stability of the e-beam energy and reduce the energy spread, injection with a tailored longitudinal density profile using a jet+cap target was investigated as in Fig. 4.6a. The mixed gas with 1 % N_2 balanced with H_e was used in both the gas jet and the capillary. The 1 J laser pulse was focused to $r_0 \simeq 21$ μm, $\Delta\tau \simeq 50$ fs and $a_0 \simeq 1.3$. The gas jet formed a high density region of ~ 300 μm at FWHM whose peak density reached $n_{\text{jet}} \sim 6$–11×10^{18}cm^{-3}. This section was coupled to a lower density plasma channel with $n_0 \sim 2$–5×10^{18}cm^{-3}.

The tailored density profile improved e-beam quality. A narrower energy spread with more stable peak energy e-beams were observed although with less charge and less injection probability. Unlike the experiment using 40 TW laser pulses presented

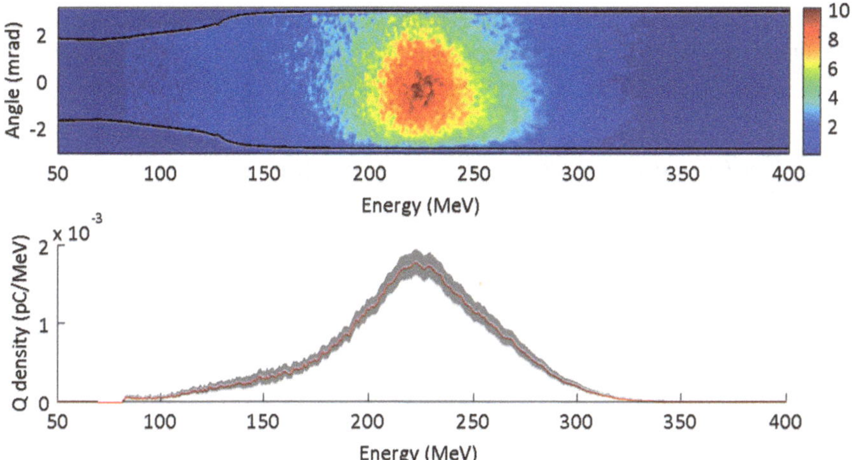

Fig. 4.13 Average spectrum of 22 e-beams produced via ionization injection using tailored plasma density in the 1st module. The charge was $Q \simeq 0.15 \pm 0.13\,\text{pC}$, $\bar{E} \simeq 221 \pm 22\,\text{MeV}$, $\theta_{\text{FWHM}} = 3.7 \pm 1.3$ ($\sigma_\theta = 1.4 \pm 0.18$) mrad, $E_{\text{peak}} \simeq 218 \pm 30\,\text{MeV}$, and $\sigma_{\Delta E/E} \simeq 16\,\%$

in Sect. 4.2.4, a_0 was not high enough to trigger injection near the negative density gradient. This was inferred from the observation that e-beam production required the low density region to be the N_2 mixed gas instead of the H_2 or H_e gas. If electrons were being injected in the negative density gradient region, e-beam production via self-trapping should be possible regardless of the gas species in the plasma channel. In addition, the laser focus had to be 1–2 mm downstream of the gas jet center, significantly downstream of the negative density gradient region, to produce e-beams. These observations suggest that electrons were being injected in the plasma channel instead of in the gas jet region. An example of an e-beam is shown in Fig. 4.13. On average, $Q \simeq 0.15 \pm 0.13\,\text{pC}$, $\bar{E} \simeq 221 \pm 22\,\text{MeV}$, $\theta_{\text{FWHM}} = 3.7 \pm 1.3$ ($\sigma_\theta = 1.4 \pm 0.18$) mrad, $E_{\text{peak}} \simeq 218 \pm 30\,\text{MeV}$, and $\sigma_{\Delta E/E} \simeq 16\,\%$ (35 MeV). Although the injection probability was 73 %, fluctuation of E_{peak} reduced to $\sigma_{E_{\text{peak}}} \simeq 14\,\%$ (31 MeV). If the energy gain from the 2nd module is $\geq$100 GeV, these e-beams will likely be sufficient. Since energy gain from the 2nd module in the initial staging experiment is expected to be less as will be discussed in Sect. 6.6, the gas jet geometry is being re-designed to further improve the stability of e-beams.

4.5 Implications for 1st Module

The staging experiment transports the e-beam without a focusing optic between the two modules while keeping the distance between the modules to a minimum ($\sim$25 mm). Section 4.5.1 discusses e-beam emittance preservation, and Sect. 4.5.2

discusses the capture condition of the e-beam in the 2nd module. These discussions will summarize what kind of e-beam properties are desirable for an ideal staged LPA.

4.5.1 Emittance Preservation Between Stages

The emittance is related to the area particles occupy in the phase space and is constant unless external forces are applied [2]. Since a small emittance implies the ability for the e-beam to be focused tightly, preserving the emittance is important for applications such as light sources and high energy colliders. The geometrical emittance of an e-beam in one transverse direction x is defined as [1, 76],

$$\epsilon_{\rm rms} \equiv \sqrt{\langle x^2\rangle\langle x'^2\rangle - \langle xx'\rangle^2}, \tag{4.1}$$

where $\langle x^2\rangle \equiv \sigma_x^2$. However, the geometrical emittance decreases as a particle is accelerated; the angle $x' = p_x/p_z$ decreases as a result of acceleration because the transverse velocity of a particle does not change while longitudinal momentum increases. Normalized emittance is defined as an invariant variable during the acceleration and expressed by $\epsilon_N = \gamma\beta\epsilon_{\rm rms}$, which was measured to be 0.1 mm-mrad for an e-beam described in Sect. 4.3.2. One source of the emittance growth in vacuum is the energy spread of e-beams. After propagation in vacuum (drift) to z, the emittance becomes [76, 77],

$$\epsilon_N(z) = \epsilon_0\left[1 + \frac{\epsilon_0^2 z^2}{\sigma_0^4\gamma_0^2}\left(\frac{\sigma_\gamma}{\gamma_0}\right)^2\right]^{1/2}. \tag{4.2}$$

Because the evolution of the transverse coordinate is $x(z) = x(0) + x'z$ where $x' = p_x/p_z$, the phase space ellipse is rotated by a different amount for a different p_z. This results in the increased emittance due to the energy spread. To simplify the equation, express the transverse size of the e-beam after propagation in vacuum to z as,

$$\sigma(z) \approx \sigma_\theta z = \epsilon_N z/\gamma\sigma_x. \tag{4.3}$$

Combining Eqs. (4.2) and (4.3) yields the change in the normalized emittance as,

$$\Delta\epsilon_N/\epsilon_N \approx \frac{\sigma(z)}{\sigma_x}\frac{\sigma_\gamma}{\gamma_0} = \frac{\sigma_\theta\, z}{\sigma_x}\frac{\sigma_\gamma}{\gamma_0}. \tag{4.4}$$

Consider the e-beam parameters discussed in Sect. 4.3.2 as an example. In this experiment, e-beam parameters were $\epsilon_N \sim 0.1$ mm-mrad, $\sigma_x = 0.1\,\mu$m, $\gamma_0 \sim 900$, and $\sigma_\gamma/\gamma_0 = 0.028$. After 25 mm propagation in vacuum which corresponds to the coupling distance between the modules for the staging experiment, the e-beam spot size is $\sigma = 25\,\mu$m. Then, the change in the normalized emittance is, $\Delta\epsilon_N/\epsilon_N \sim 7$.

If the e-beam energy is lower, and σ_γ/γ_0 larger as was observed in many e-beams produced via self-trapping and ionization of nitrogen atoms, the emittance will grow faster. However, if the correlated energy spread were removed and slice energy spread were <0.5 % as measured by OTR, the emittance growth will be a smaller effect, $\Delta\epsilon_N/\epsilon_N \sim 1$. This sensitivity to the e-beam spot size, divergence, energy and energy spread has to be considered for optimizing future staged LPAs.

One way to mitigate emittance degradation between the modules, in addition to reducing e-beam energy spread, is by expanding the e-beam spot size and reducing the divergence through tailoring the focusing force at the exit of the plasma channel. The transverse e-beam size in the plasma wakefield is determined by the emittance and the transverse focusing force from the plasma waves. The e-beam radius in plasma evolves as [73],

$$\frac{d^2\sigma_x}{dz^2} = \left(\frac{\epsilon_N}{\gamma}\right)^2 \frac{1}{\sigma_x^3} - k_\beta^2\sigma_x, \tag{4.5}$$

where $k_\beta = k_p/\sqrt{2\gamma}$ from the betatron oscillation. By gradually reducing the focusing force through decreasing plasma density, the e-beam expands in size and reduces in divergence. For this effect to be significant, the plasma density decrease must take place in a distance that is long compared to the betatron wavelength ($L \gg \lambda_\beta \approx \lambda_p\sqrt{2\gamma}$) [73].

Plasma density downramp effects on the emittance for a staged LPA were investigated with simulations by Vay et al. [78]. For this study, laser spot size was $r_0 = 9\,\mu$m and $a_0 = 1$. Each LPA module was 1 mm long and the coupling length, the distance between the modules, was 0.25 mm. The on-axis density in each module was $n_0 = 10^{19}$cm^{-3}, and the longitudinal plasma density profiles are shown in Fig. 4.14a. Simulated emittance growths as functions of propagation distance for the two density profiles are shown in Fig. 4.14b. When each stage has a plasma density downramp of $\sim 230\,\mu$m $\approx 23\lambda_p$ [dot-dashed blue curve in Fig. 4.14a], which corresponds to

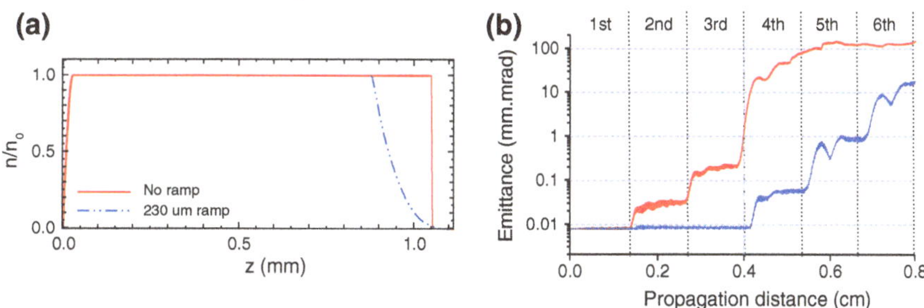

Fig. 4.14 Simulated effects of the density downramp on the e-beam emittance for multiple LPA modules. **a** Longitudinal plasma density profile (n/n_0) in each module. A profile without density downramp (*red*) and with 230 μm long downramp (*dot-dashed blue*). **b** The emittance as a function of propagation distance over six modules. Figures from from [79]

approximately λ_β for $\gamma = 260$, the emittance is preserved for three modules instead of one module as in the case without the density downramp [red curve in Fig. 4.14b]. For our experiments with $n_0 = 10^{18}\text{cm}^{-3}$, the equivalent λ_β for $\gamma = 260$ is $\sim 800\,\mu\text{m}$. For higher energy beams, λ_β becomes longer ($\lambda_\beta \simeq 1.5\,\text{mm}$ for $\gamma = 1000$) and requires a longer ramp. However, since plasma channels are much longer in experiments (33 mm), the use of a density downramp on the order of a few millimeters is feasible. In addition, the emittance growth is proportional to γ^{-1}, and the growth will be less significant for the higher energy beams. This study showed that by adjusting the lengths and the profile of the density downramp appropriately for e-beams, the emittance preservation can be tuned between many modules.

Another effect that may be able to mitigate the emittance growth is a decrease in a_0 at the capillary exit due to the mismatched laser guiding in the plasma channel, although this effect requires further investigation. As was discussed in Sect. 2.3.2, the laser spot size can oscillate in the plasma channel when the laser and plasma channel parameters do not allow for perfect cancellation of the laser diffraction. Often, this mismatched guiding results in decreasing a_0 as the laser exits the capillary. Figure. 4.15a shows the simulated laser a_0 and longitudinal plasma density n/n_0 as a function of propagation distance for parameters used in an experiment. The simulated parameters are $r_0 = 18\,\mu\text{m}$, $\Delta\tau = 75\,\text{fs}$, $a_0 = 1.2$, $n_0 = 1.5 \times 10^{18}\text{cm}^{-3}$, $r_\text{m} = 41\,\mu\text{m}$ and the laser was focused 4.8 mm downstream of the channel entrance. Due to the mismatched guiding, the a_0 oscillates in the plasma. This results in decreasing a_0 (and increasing laser spot size) at the capillary exit, exciting lower wake amplitudes. Figure. 4.15b shows the transverse electric field for various longitudinal positions near the capillary exit. The transverse focusing forces decrease as the e-beam approaches the capillary exit due to the decreasing a_0 of the driving laser. This decreasing focusing force may allow the e-beam spot size to expand and reduce the divergence as it exits the capillary. Although further study including the

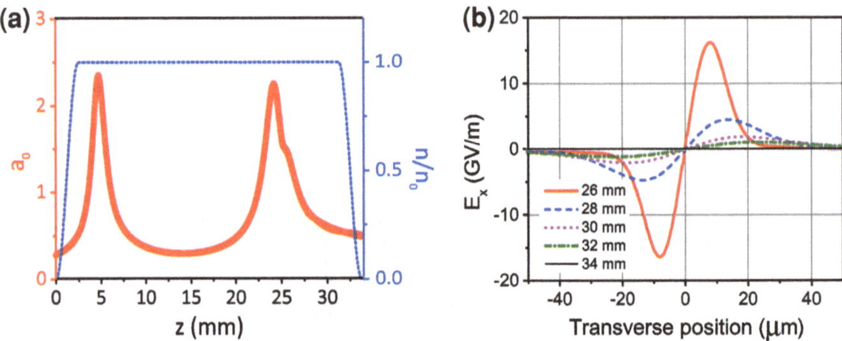

Fig. 4.15 Simulations to study the effects of mismatched laser guiding on the transverse focusing force near the exit of the capillary. **a** The laser a_0 and n/n_0 within a stage. **b** Fields at various longitudinal locations near the capillary exit are shown. The e-beam traveling at the speed of light will experience decreasing transverse focusing force as it exits the plasma channel

effect of dephasing is required, this may expand the e-beam size and mitigate the emittance growth in the drift space between each LPA module.

4.5.2 Electron Beam Capturing Conditions at 2nd Module

To minimize betatron radiation and preserve the emittance of accelerating e-beam in the 2nd module, the e-beam size and emittance must be matched to the plasma wave. First, an estimate of electron capturing efficiency will be discussed. Then an estimate of matching efficiency will be discussed.

One way to estimate the fraction of electrons captured in the plasma waves in the 2nd module is to compare the size of the e-beam with that of the plasma wave at the entrance of the 2nd module. The size of the e-beam can be estimated using Eq. (4.3). The size of the plasma wave is roughly the transverse size of the driving laser, r_0. Then, based on the geometry, the fraction of e-beam that can be captured in the plasma wave is [76],

$$f(z) = 1 - \exp\left(-\frac{r_0(z)^2}{2\sigma(z)^2}\right). \tag{4.6}$$

The calculated e-beam sizes at the 2nd module for various e-beam divergences are shown in Fig. 4.16a. A 1 mrad divergence e-beam will be $\sigma_x \approx 25\,\mu\text{m}$ after 25 mm of vacuum propagation. If the drive laser spot size $r_0 \approx 20\,\mu\text{m}$, approximately 27 % of electrons will fall into the wakefield. These electrons may be able to post-accelerate in the 2nd module.

Without focusing e-beams between the stages, e-beams will likely undergo mismatched propagation in the 2nd module. The large plasma focusing forces can lead to strong betatron oscillations, resulting in the loss of energy and an increase in

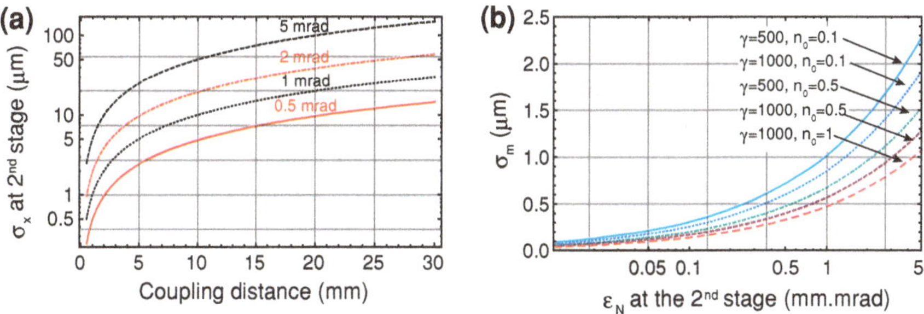

Fig. 4.16 **a** Calculated electron beam spot size at the entrance of the 2nd module for the divergence observed in LPA experiments (0.5-5 mrad). **b** Calculated matched e-beam sizes as a function of the emittance are shown for several e-beam energies, γ, and plasma density, n_0. The n_0 are in units of 10^{18}cm^{-3}

energy spread due to synchrotron (betatron) radiation. When e-beam diffraction is perfectly compensated by the transverse force, the e-beam is said to be matched to the plasma wave and propagates at a constant radius σ_x. The condition for a matched e-beam is,

$$\sigma_m = (\epsilon_N/\gamma k_\beta)^{1/2}. \tag{4.7}$$

An example of the matched e-beam production was the X-ray spectroscopy-based emittance measurement experiment presented in Sect. 4.3.2 [68]. In the experiment, $n_0 = 5 \times 10^{18}\,\text{cm}^{-3}$, $\gamma = 1,000$, $\epsilon_N = 0.1$ mm-mrad yielding $\lambda_\beta = 0.66$ mm, and $\sigma_m = 0.1\,\mu$m, which was the value measured from the X-ray spectrum. For the 2nd module in the staging experiment, e-beam will likely be mismatched. The calculated matched e-beam sizes as a function of the emittance are shown in Fig. 4.16b for several e-beam energies. With an emittance of $\epsilon_N = 0.01$–5 mm-mrad, matched e-beam size is $\sigma_\text{m} < 2.5\,\mu$m. For $n_0 = 10^{18}\text{cm}^{-3}$, $\gamma = 500$, and $\epsilon_N = 1$ mm-mrad, the matched spot size is $\sigma_\text{m} \sim 1\,\mu$m. Assuming e-beam divergence of ~1 mrad, frequently observed during the injection experiments discussed in Sect. 4.2, the e-beam spot size after 25 mm of vacuum propagation is $\sigma_x \simeq 25\,\mu$m and is significantly greater than σ_m. Without a decrease in the e-beam size, e-beams will likely undergo mismatched propagation in the 2nd module. One way to better match e-beam spot size into the plasma wave is to use quadrupole magnets to focus the e-beams between the modules. Using a combination of focusing and defocusing quadrupole magnets, focusing can be obtained for both horizontal and vertical axis.

4.6 Summary and Conclusions

Experiments on e-beam production using 40 TW TREX laser pulses in the undulator beamline were presented. At this power level, e-beam production can occur through self-trapping and/or ionization in a capillary. Both injection mechanisms resulted in e-beams with large energy spread and large energy fluctuations. High quality, stable energy e-beams were demonstrated when a capillary waveguide with a gas jet embedded in the upstream region to tailor the longitudinal plasma density profile was employed. Electron beams with peak energy $\simeq$ 340 MeV with fluctuation and spread of only $\sigma_E \simeq 1.9\,\%$ and $\sigma_{\Delta E/E} \simeq 5\,\%$ were observed at ~100 % probability. These e-beams would be suitable for the staging experiment.

Experiments on characterization of self-trapped e-beams in the undulator beamline were presented. For successful coupling and the post-acceleration of e-beams in a staged LPA, understanding e-beam properties are critical. Observation of a laser induced femtosecond structure using COTR in broad energy spread e-beams was reported. The coherent enhancement of OTR suggested an energy slice spread of $\Delta\gamma/\gamma \leq 0.5\,\%$ from large integrated energy spread e-beams. In addition, the emittance of e-beams was estimated using e-beam size and divergence measurements based on X-ray and e-beam spectra. The measured normalized e-beam emittance ranged from 0.1 to 1 mm-mrad.

The initial experimental results on e-beam production in the staging setup were presented. Using a combination of a tailored density profile and ionization of nitrogen atoms, e-beams with peak energy $\simeq$218 MeV with fluctuation and spread $\sigma_E \simeq 14\,\%$ and $\sigma_{\Delta E/E} \simeq 16\,\%$ were produced. These e-beams will likely be sufficient for the staging experiment if the energy gain in the 2nd module is $\geq$100 MeV.

Preservation of e-beam qualities between the modules was discussed. The emittance of e-beams from LPAs can degrade significantly if the e-beam spot size is small, divergence is large, and energy spread is large. As the e-beam increases its energy, the emittance growth will reduce. However, in the first module where e-beams are relatively lower energy, the emittance preservation must be considered for designing staged LPAs. This can be achieved through decreasing the focusing force via plasma density downramp, possibly expanding the laser spot size as is the case for mismatched laser guiding or by obtaining a higher energy in the 1st module. To capture the e-beam and post-accelerate efficiently in the 2nd module, the transverse size of the e-beam, σ_x, and the matched spot size of the plasma wave, σ_m, should be matched at the entrance of the 2nd module. This may be achieved when the e-beam is focused between the modules.

In a staged LPA, production of a high quality e-beam is more critical than ever since the e-beam must be transferred to the sequential module and accelerated in a controlled manner. Since various injection techniques using colliding pulses, negative density gradients and ionization are being actively investigated in the LPA community, improvements on e-beam properties are anticipated [20, 23, 24, 44, 45]. These improvements on e-beam quality and control are exciting topics for future staged LPAs.

Chapter 5
Plasma Mirror

5.1 Introduction

In this chapter, development and characterization of a plasma mirror are presented. The theoretical framework of the plasma mirror is discussed in Sect. 5.2. A tape-drive based plasma mirror was developed and its properties such as reflectivity, optimum fluence, reflected mode quality and reflected laser pulse pointing fluctuations were investigated. These experimental results are presented in Sect. 5.3.

The overall accelerating gradient of an LPA-based high energy accelerator will be strongly influenced by the coupling distance (L_c) between each module. Figure 5.1 shows the total accelerator length for an ideal LPA producing 500 GeV electron beams as a function of operating plasma density n_0. Diffraction and dephasing are assumed to be controlled, and the electron energy gain in each module is the ideal energy gain limited by pump depletion as was discussed in Sect. 2.5. The assumed drive laser is $\lambda = 0.8\,\mu$m and $a_0 = 1.5$. The plot shows that the total accelerator length quadruples (overall acceleration gradient is reduced by factor of 4) when $L_c \sim 50$ cm compared to $L_c \sim 10$ cm at a plasma density of $n_0 = 10^{18}\,\text{cm}^{-3}$.

A plasma mirror is critical for achieving a high overall accelerating gradient because use of a conventional optic for drive laser coupling results in a large L_c. The fluence (energy/area) of the drive laser near the laser focus is 100–1,000 J/cm^2, far above the damage thresholds of any conventional optics (which are 1–3 J/cm^2) [80]. A schematic of the coupling distance using a conventional mirror is shown in Fig. 5.2a. If a conventional mirror were to be used, the mirror must be placed 50–200 cm upstream of the laser focus (where the spot size is large) in the staging experiment. This large L_c requires electron transport optics between the injection and acceleration stages, and significantly reduces the overall accelerating gradient of the LPA. The use of a plasma mirror is one way to overcome the limitation of conventional mirrors and retain the high accelerating gradient in staged LPAs. A schematic is shown in Fig. 5.2b [57]. The plasma mirror is triggered on the surface of a solid target at $\sim$6 J/cm^2 and can be operated at a few hundred J/cm^2, allowing placement near focus [81]. A schematic of a plasma mirror is illustrated in Fig. 5.2c and its formation

S. Shiraishi, *Investigation of Staged Laser-Plasma Acceleration*,
Springer Theses, DOI: 10.1007/978-3-319-08569-2_5

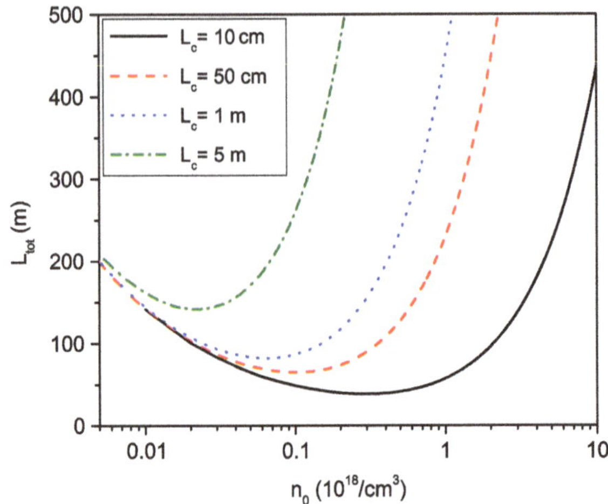

Fig. 5.1 Theoretical accelerator lengths as a function of plasma density for various coupling lengths for 500 GeV final electron energy. Driving laser is assumed to have $a_0 = 1.5$ and $\lambda = 0.8\,\mu$m. Diffraction and dephasing are assumed to be controlled and the length of each LPA module is limited by pump depletion

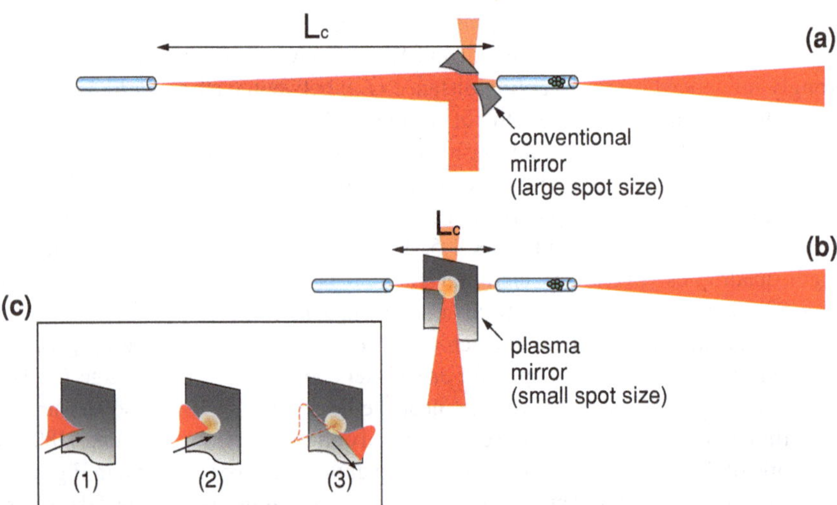

Fig. 5.2 Schematics of plasma mirrors. The coupling lengths L_c will be large when a conventional mirror is used to merge laser pulse 2 onto the electron beamline (**a**). The L_c will be much shorter when a plasma mirror is used (**b**). Triggering of the plasma mirror is illustrated in (**c**): (*1*) a laser pulse strikes the surface of solid target, (*2*) the temporal leading edge of the laser pulse ionizes the surface of the target, forming a plasma, (*3*) when the plasma density reaches the critical density, the rest of the laser pulse is reflected

can be described in three steps: (1) a laser pulse strikes the surface of solid target, (2) the leading edge of the laser pulse ionizes the surface of the target forming a plasma, and (3) when the plasma density reaches the critical density $n_c = \omega_L^2 m_e / 4\pi e^2$, the rest of the laser pulse is reflected. The mirror is disposable, and a fresh surface of the solid target is provided for each laser shot. The understanding and use of the plasma mirror is crucial for minimizing L_c and achieving a high overall average accelerating gradient in staged LPAs [27, 58].

5.2 Theoretical Framework of Plasma Mirror

A plasma mirror is triggered when the laser field ionizes the surface and forms a layer of plasma. The laser field continues to ionize the target until the electron density reaches the critical density, n_c, at which time the surface of the plasma behaves as a mirror and reflects the rest of the pulse. The critical density can be seen as the density where the index of refraction becomes imaginary. The index of refraction of a plasma is defined as $n_R = (1 - \omega_p^2/\omega_L^2)^{1/2}$ where $\omega_p = (4\pi e^2 n_e / m_e)^{1/2}$, ω_L is the angular frequency of the laser pulse, and n_e is the electron density. The laser pulse is transmitted when $n_R > 0$ and reflected when n_R becomes imaginary at $\omega_p = \omega_L$. This occurs at the critical density and $n_c \sim 2 \times 10^{21}\,\text{cm}^{-3}$ for $\lambda = 800\,\text{nm}$ [5].

For a femtosecond laser pulse, the surface quality of the plasma mirror is similar to that of the original target before ionization because the pulse is reflected faster than the time scale for hydrodynamic expansion. The plasma pressure from the heating of the target causes the expansion at roughly the speed of sound (in cgs units) [12]:

$$c_s = \left(\frac{Z k_B T_e}{m_i}\right)^{1/2} \simeq 3.1 \times 10^7 \left(\frac{T_e}{\text{keV}}\right)^{1/2} \left(\frac{Z}{A}\right)^{1/2} \text{cm s}^{-1}, \tag{5.1}$$

where Z is the ionization degree, k_B the Boltzmann constant, T_e the electron temperature, m_i the ion mass, and A is the atomic mass [81]. Typically, the plasma of $T_e \sim 100\,\text{eV}$ expands ~10–100 nm during the 100 fs laser pulse impinging on a solid target [12]. Since the laser spot size on target, 100–500 μm, is much larger than the expansion extent, the original surface quality of the tape was critical for the experiments.

The plasma expands at roughly the ion acoustic speed and can cause spectral blueshifting of the reflected pulse. A schematic of the plasma expansion relative to the impinging laser pulse is shown in Fig. 5.3. This expansion moves the reflective surface for the laser pulse and the Doppler shifts compresses the laser field. Spectral blueshifts from plasma expansion were reported in previous experiments [82, 83].

The reflectivity of the plasma mirror depends on the polarization of the impinging laser field. Since a s-polarized electromagnetic (EM) wave (**E** perpendicular to the incident plane) is purely transverse, there is no coupling with the electrostatic modes of the plasma density gradient ($\nabla \cdot \mathbf{E} = 0$). For an s-polarized EM wave traveling in the z-direction, $\mathbf{E} = (0, E_y, 0)\, e^{ikx \sin\theta}$, the Helmholtz equation is expressed as [81]

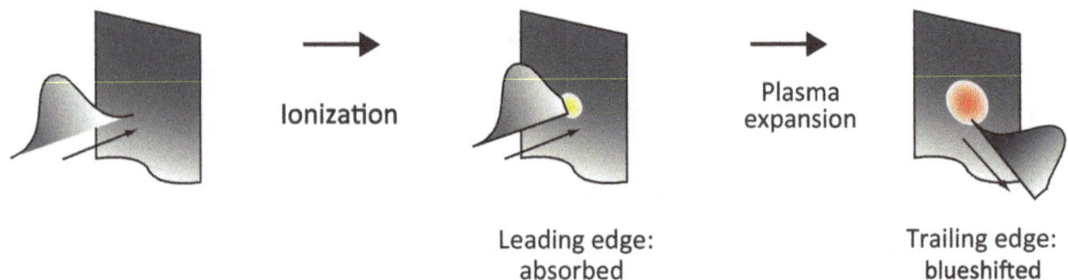

Fig. 5.3 Schematic of plasma expansion effect on a chirped laser pulse. The *leading edge* of the laser pulse is absorbed for the ionization. The *trailing edge* can experience spectral *blueshift* due to the Doppler effect from the expanding *plasma surface* approaching the incident laser pulse. The plasma mirror expands roughly at the ion acoustic speed

$$\frac{\partial^2 E}{\partial z^2} + k^2 \left(\varepsilon - \sin^2\theta\right) E = 0. \tag{5.2}$$

On the other hand, a p-polarized EM wave, $\mathbf{E} = (E_x, 0, E_z)$, has a component in the plane of the plasma density gradient ($\nabla \cdot \mathbf{E} \neq 0$). The E_z component along the plasma density gradient drives an electrostatic charge density oscillation in the plasma mirror at $\omega = \omega_\mathrm{p}$. This transfer of energy from the EM (laser) to electrostatic (plasma) wave is called resonance absorption [12, 84]. The resonance absorption reduces the reflectivity of the laser reflected off the plasma mirror [81, 85].

The maximum reflectivity of a plasma mirror can be estimated to first order with the Fresnel equation for s-polarized light. When an intense EM wave impinges on an overdense plasma, the EM field forms a standing wave in front of the target along with an evanescent component penetrating into the overdense plasma region. The scale length of the plasma density is L/λ, where $L = c_\mathrm{s}\tau_L \sim 10\,\mathrm{nm} \sim 0.01\lambda$ is the plasma expansion for the femtosecond duration of our laser pulse τ_L. In the limit of $L/\lambda \to 0$, approximately satisfied in this case, the Fresnel equations of metal optics are recovered from these Helmholtz equations in Eq. (5.2). In addition, a 1-D limit is considered since the transverse laser spot size is much greater than the plasma expansion scale ($L \ll r_0$) assuming the plasma expands isothermally. Consider a Heaviside step function for the plasma density, $n_0(z) = n_0\Theta(z)$, and the dielectric constant [86],

$$\varepsilon \equiv k^2c^2/\omega^2 = 1 - \frac{\omega_\mathrm{p}^2}{\omega^2(1 + i\nu_{ei}/\omega)}, \tag{5.3}$$

where ν_{ei} is the electron-ion collision frequency expressed as,

$$\nu_{ei} = \frac{4(2\pi)^{1/2}}{3} \frac{n_\mathrm{e} Z e^4}{m_\mathrm{e}^{1/2}(k_\mathrm{B}T_\mathrm{e})^{3/2}} \ln\Lambda \simeq 2.9 \times 10^{-6} Z n_\mathrm{e} T_\mathrm{e}^{-3/2} \ln\Lambda\ \mathrm{s}^{-1}, \tag{5.4}$$

and $\ln\Lambda$ is the Coulomb logarithm which is roughly 1–10. For the s-polarized wave, a solution to Eq. (5.2) is [86],

$$E = \begin{cases} E_0 \exp[i(kz - \omega t)], & \text{incident field } (z < 0) \\ \sqrt{R} E_0 \exp[i(-kz - \omega t)], & \text{reflected field } (z < 0) \\ E_0 \exp(-z/l_s), & \text{transmitted field } (z \geq 0), \end{cases} \tag{5.5}$$

where R is the reflectivity and l_s the skin-depth. In the overdense region ($z \geq 0$), the field is evanescent. Neglecting collisions in the dielectric constant, the skin depth is [86],

$$l_s = \frac{c}{\omega_\mathrm{p}} \left(1 - \frac{w^2}{\omega_\mathrm{p}^2} \cos^2 \theta\right)^{1/2} \simeq \frac{c}{\omega_\mathrm{p}}, \tag{5.6}$$

where the second equality on the right resulted from the assumption $n_0/n_c \gg 1$. At this limit, reflectivity is derived with Fresnel equations:

$$R_{\mathrm{s-pol}} = \left| \frac{\cos\theta - \sqrt{\varepsilon - \sin^2\theta}}{\cos\theta + \sqrt{\varepsilon - \sin^2\theta}} \right|^2 . \tag{5.7}$$

In reality, the dielectric constant is a function of time, and a more rigorous calculation of the plasma mirror reflectivity can be found in Ref. [81]. The varying dielectric constant is due to the penetration depth of the laser field in the material which decreases with time. Early in the interaction when the laser impinges on the target, the electrons in the material are homogeneously excited on a micron length scale. As the laser intensity increases, the excitation of the electrons increases, which in turn reduces the laser field that penetrates into the material. As a result, electron density near the surface increases more with time than it does deeper in the material. When the density exceeds the critical density, this effect is even stronger since the laser does not penetrate further than the skin depth.

In brief, the plasma mirror is a dynamic disposable mirror with its reflective properties depending on laser fluence, pulse duration, polarization, angle of incidence and the dielectric constant of the material ε which is a function of electron-ion collision frequency. In the next section, the characterization of a tape-drive based plasma mirror using VHS tapes are presented.

5.3 Experimental Configuration and Results

Use of the plasma mirror in LPA experiments required investigation of thin targets that had never been used as plasma mirrors. Plasma mirrors are conventionally formed on thick optical quality solid disks and substrates such as quartz to improve temporal contrast of laser pulses. They have demonstrated reflectivity of 70–80 % and high quality reflected laser profiles [81, 83, 87, 88]. However, the staging experiment imposed two additional requirements. First, e-beams have to propagate through the

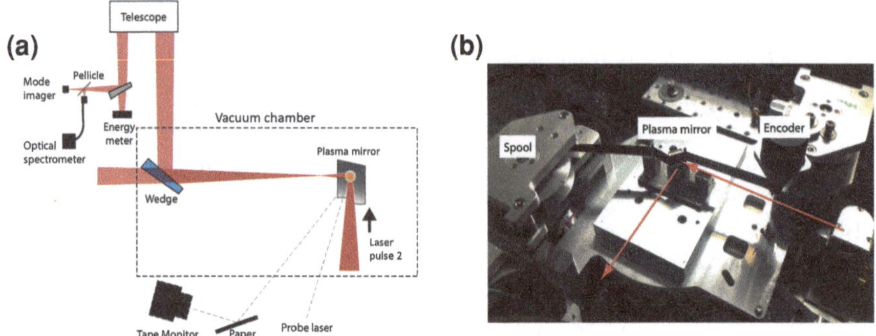

Fig. 5.4 Experimental setup for the plasma mirror characterization. **a** Laser is focused downstream of the plasma mirror target. A reflected laser pulse is attenuated with an uncoated wedge and focused with a telescope. The laser energy, mode, and optical spectrum are measured after the telescope. A probe laser is shone on the VHS tape and reflected profile is recorded with a CCD camera to monitor the tape surface quality. **b** A photo of the tape drive for plasma mirror. VHS tape is shown and the laser pulse 2 path is indicated with *red arrows*

plasma mirror, which is not possible when the plasma mirror is formed on a thick substrate. The plasma mirror target has to be very thin to minimize the influence on e-beams. Second, many thousands of uninterrupted laser shots are necessary for LPA experiments. The number of shots that can be applied on the substrate is limited by the surface area since each laser pulse must strike a fresh surface. To overcome these constraints, a tape-drive based plasma mirror was developed. A long thin tape is spooled through the tape-drive, and thousands of shots can be applied without interrupting the experiments.

The experiment was performed using laser pulse 2 in the staging beamline which used s-polarized laser pulses. The experimental setup is shown in Fig. 5.4a. The total laser energy on target was ~650 mJ, input laser focus spot size was $r_0 = 21\,\mu\text{m}$, and pulse duration ranged from 55–140 fs at FWHM. The laser fluence on the target was adjusted by moving the laser focus position, 5–30 mm downstream, with respect to the tape. Assuming a perfect Gaussian profile, these focus positions corresponded to an average fluence of 100–3,500 J/cm^2. The reflected laser mode, energy and optical spectra were recorded for each shot. A photo of the tape drive in the vacuum chamber is shown in Fig. 5.4b. The laser pulse propagated from the lower right of the image and struck the surface of the tape, forming a plasma mirror.

Surface roughness of available plastic strips were measured using a Micromap interferometer at the Optical Metrology Laboratory at LBNL and examples are shown in Fig. 5.5 [89]. Both plastics, Mylar and VHS, shown in Fig. 5.5 are thought to be polyester film (also referred to as PET), but the different process of fabrication resulted in different qualities of the plastics. While the exact composition of the VHS tapes are not disclosed by the manufacturer, most likely the VHS tape consists of an iron-oxide layer on a plastic film, and the plastic side was used in our experiments [90]. The rms variation of the Mylar tape surface was 68 nm and that of the

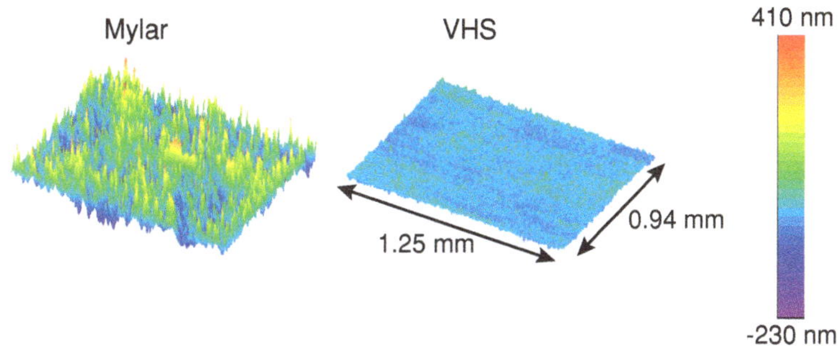

Fig. 5.5 Measurements of surface qualities for Mylar and VHS tapes using a Micromap interferometer [89]. The field of view is 1.25 mm × 0.94 mm. The rms roughness of the Mylar surface is 68 nm and that of the VHS surface is 16 nm. Both plastics are 25.4 μm thick

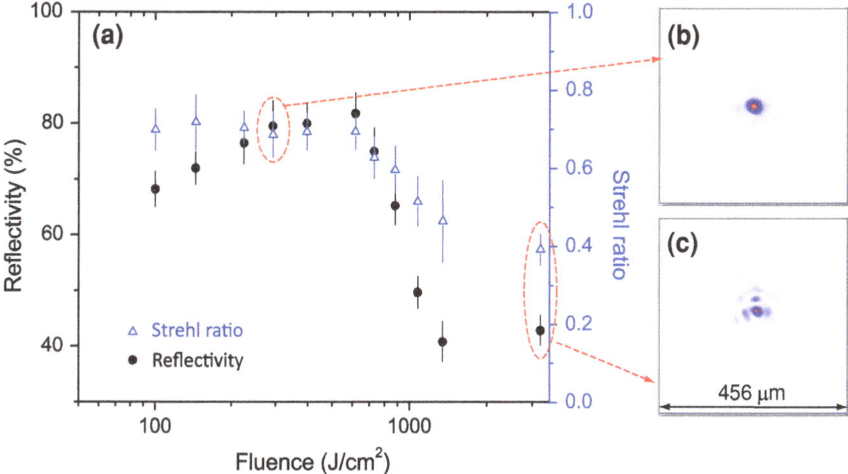

Fig. 5.6 **a** Plasma mirror reflectivity and Strehl ratio of reflected pulses as a function of average fluence on the tape. The fluence is calculated assuming an ideal Gaussian profile. **b** and **c** Examples of reflected laser mode images at focus for 300 J/cm^2 and 3, 500 J/cm^2. Counts on (**c**) are multiplied by factor of two

VHS tape surface was 16 nm. This measurement indicated that the VHS tape was the most promising candidate for the plasma mirror because the tape had a better surface quality and was only 25 μm thick.

Fluence dependent reflectivity and reflected laser mode quality were observed using the VHS tape. Measured reflectivity and laser focus quality quantified by the Strehl ratio as functions of fluence on target are shown in Fig. 5.6a. The fluence of ~300–700 J/cm^2 on the target was the optimum for both reflectivity and reflected mode quality. The maximum reflectivity was ~80 % and the best Strehl ratio was 0.7 which indicated that the mode quality was as good as incident pulses as seen in Fig. 5.6b. When the fluence was lower than optimum (<250 J/cm^2), the plasma mirror triggered too late with respect to the arrival of the main pulse, transmitting a larger fraction of laser energy. In this region, the reflectivity was low but with a high Strehl ratio. When the fluence was higher than optimum (>700 J/cm^2), the plasma seemed to be heated too much. When the plasma temperature T_e is high, the plasma expansion scale and electron-ion collision frequency increases, resulting in lower reflectivity and increased surface distortion. In this region, both the measured reflectivity and the Strehl ratio were low as seen in Fig. 5.6c. These measurements demonstrated that a sufficient plasma mirror performance can be achieved when laser fluence on the tape is ~300–700 J/cm^2.

The theoretical reflectivity of a plasma mirror achievable from the VHS tape was estimated by the Fresnel equation (5.7), to compare with the experimental result. The maximum plasma density on the plastic surface of the VHS tape was measured using the high-harmonic generation at the LOASIS facility [91]. In this experiment, harmonics up to 16th order were observed from the plastic surface of the VHS tape, implying a maximum plasma density of $n_e \sim 4.35 \times 10^{23}$ cm^{-3} on the tape [87]. This density results in a skin depth $l_s \sim 8$ nm and collision frequency $\nu_{ei} \sim 10^{14}$ Hz for $T_e \sim 100$ eV. With these parameters, the theoretical maximum reflectivity for the s-polarized pulse with an incident angle of 45° is 84 %. Considering the Fresnel equation is an approximation that does not take into account the time varying dielectric constant, the experimentally observed reflectivity of 80 % was reasonably close to optimum.

Measured optical spectra on chirped pulses showed the spectral blueshift from the plasma expansion. Figure 5.7 shows measured spectra for two different chirped pulses impinging on a plasma mirror at a fluence of 500 J/cm^2. The black curves are before the plasma mirror and blue dot-dashed curves are after the plasma mirror. Each line is an average of ~5 shots and the standard deviation is shown with grey

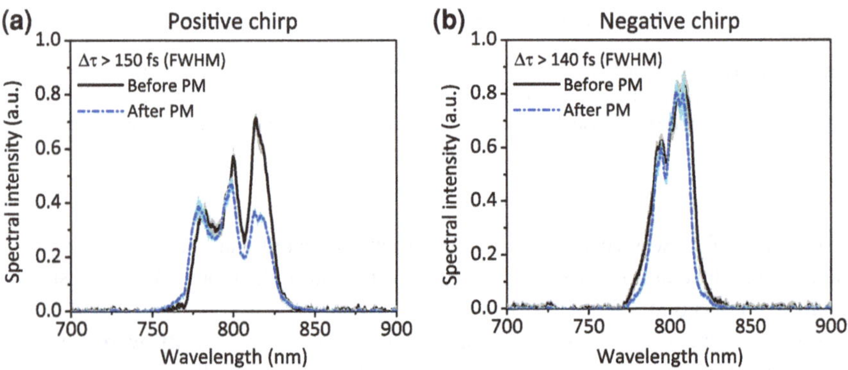

Fig. 5.7 **a** Measured spectra before and after the plasma mirror for a positively chirped pulses showing spectral blueshift. **b** Measured spectra before and after the plasma mirror for a negatively chirped pulses showing spectral narrowing. The *black curves* are before the plasma mirror and *blue dot-dashed curves* are after the plasma mirror. Each line is an average of ~5 shots and the standard deviation is shown with *grey* and *light blue lines*

and light blue lines. The input spectra were different for the two cases because of the different amount of self-phase modulation experienced in the beam splitter which was placed upstream of the plasma mirror. This effect will be discussed further in Chap. 6. As was discussed in Fig. 5.3, the leading edge of the laser pulse is absorbed and the trailing edge of the laser pulse is spectrally blueshifted from the plasma expansion. For a positively chirped pulse (wavelength decreases with time), the longer wavelengths are absorbed while the shorter wavelengths are blueshifted. As a result, the spectra reflected off the plasma mirror shifts towards shorter wavelengths as seen in Fig. 5.7a. For a negatively chirped pulse (wavelength increases with time), the shorter wavelengths are absorbed for ionization and the longer wavelengths are blueshifted, resulting in a spectrally narrower laser pulse as seen in Fig. 5.7b. Taking the $\Delta\lambda = -2$ nm observed in Fig. 5.7a, the plasma expansion speed was estimated to be $c_s \sim 8 \times 10^7$ cm/s. It should be noted that the spectral blueshift can also be caused by the increasing plasma density during the ionization process on the tape. Laser field in the tape propagates in an increasing plasma density and can undergo frequency up-shift because phase velocity of a laser field depends on the plasma density. This effect would decrease the estimated c_s from the Doppler effect. Therefore, the c_s measurement supports the importance of the tape surface quality before ionization for our experiment since $c_s\Delta\tau < 1\,\mu\text{m} \ll r_0$.

A technical challenge with the plasma mirror used in the staging experiment was establishing stable reflected laser pulse pointing. If the drive laser pointing were fluctuating, the excited wakefield would not be co-linear with the e-beam propagation axis, resulting in significant betatron oscillations and/or e-beam loss. Causes of fluctuations in the reflected laser pulse pointing included inconsistent tape surface positioning and damaged tape surfaces. Movement of the tape to provide a fresh surface for every laser shot increased the chance the tape was not placed reproducibly. To keep the surface as flat as possible, a constant tension was applied to the tape which occasionally over-stretched the tape. The surface may also have been scratched or damaged from dust and debris. To minimize laser fluctuations from these effects, the surface quality of the tape prior to plasma mirror formation was monitored during the experiments and the layout is shown in Fig. 5.4a. A low power continuous wave laser was shone on the tape and the reflected laser mode was monitored by a CCD camera. The profile of this probe laser indicated if the tape surface was distorted.

Reflected laser pointing sufficient for the staging experiment was achieved when only a good surface was used. Figure 5.8a shows the measured reflected pulse centroid 18 mm downstream of the plasma mirror near where the entrance of the 2nd module will be placed. Data points from tape surfaces deemed not acceptable based on the tape monitor camera were filtered out in this plot. Since plasma channels of $r_m \sim 30$–$40\,\mu$m are used in the staging experiment, these reflected laser pulses will propagate through the channel. Figure 5.8 shows images of the tape monitor camera for a good surface (b) and a poor surface (d). Respective reflected mode images are shown as Fig. 5.8c, e. The reflected pulse from a good surface (c) has Strehl ratio of 0.7 and is positioned near the center of the image. The pulse reflected off a poor surface (e)

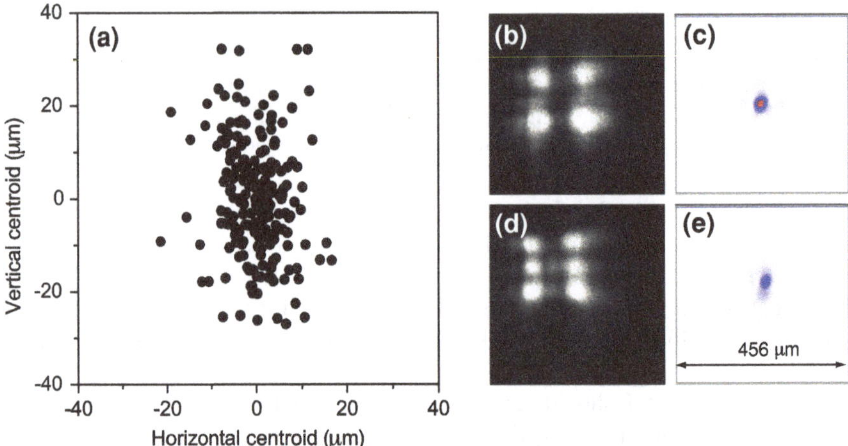

Fig. 5.8 Laser pulse pointing measurements reflected off the plasma mirror. **a** Measured reflected pulse centroid at 18 mm downstream of the plasma mirror. The standard deviations on horizontal and vertical fluctuation are 5.8 and 12.3 μm. Typically, plasma channels of $r_m \sim$ 30–40 μm are used for the 2nd module, so the reflected pulses shown here will propagate through the channel. **b** Tape monitor probe laser indicating a good VHS tape surface. **c** Reflected pulse off the surface shown in (**b**). **d** Tape monitor probe laser indicating a poor surface. **e** Reflected pulse off the surface shown in (**d**)

has Strehl ratio of 0.6 and is positioned higher on the vertical axis. This is probably because the VHS tape was over-stretched. In LPA experiments, laser pulses were fired only on a good surface. The tape monitoring diagnostic reduced laser pulse pointing fluctuation to a level sufficient for the staging experiment.

5.4 Electron Beam Interaction with Plasma Mirror

The dynamical coupling of the e-beam with the plasma mirror is estimated to be small. Assuming an e-beam size of 10 μm both transversely and longitudinally with 10 pC of charge at the plasma mirror, the electron density in the e-beam is $n_b \sim 6 \times 10^{19}\,\mathrm{cm}^{-3}$. As was discussed in Sect. 5.3, the electron density in the plasma mirror was estimated to be $n_e \sim 4 \times 10^{23}\,\mathrm{cm}^{-3}$ [91]. Since the electron density in the e-beam is much smaller than that in the plasma, $n_b/n_e < 10^{-4}$, and the plasma wavelength is much shorter than the e-beam length ($\lambda_p \sim 0.05\,\mu\mathrm{m} \ll 10\,\mu\mathrm{m}$), the coupling of the e-beam with the plasma wave will not be the dominant effect [76].

The effect of electron-ion scattering (i.e. small angle scattering) on e-beams propagating through the plastic tape is estimated. The rms divergence of an e-beam propagating through a thin solid material is calculated using [92],

$$\theta_{\rm rms} = \frac{17.5}{\gamma mc^2}\sqrt{\frac{z}{X_0}\frac{9}{8}}\left[1+\frac{1}{9}\log_{10}\left(\frac{z}{X_0}\right)\right], \tag{5.8}$$

where z is the propagation distance and $X_0 = 28.5$ cm is the radiation length of a polyester film. When a $\sim$ 511 MeV ($\gamma = 1{,}000$) e-beam propagates through a tape of 35 μm thick (25 μm thick tape placed at 45°), the scattering angle is $\theta_{\rm rms} \simeq$ 247 μmrad. This scattering is proportional to γ^{-1} and reduces for higher energy e-beams.

In addition to the scattering in the un-ionized tape, e-beams in a staged LPA also propagate through a plasma mirror. Since electrons are not bounded to atoms in a plasma, electron-ion scattering can be greater than that in un-ionized solid. The plasma mirror is a dense sheet of plasma of a few skin-depths thick formed on the surface of the plastic tape. A change of the rms divergence through Coulomb scattering of e-beam with background ions in the plasma mirror can be estimated with [28],

$$\frac{\mathrm{d}\langle\theta^2\rangle}{\mathrm{d}z} = \frac{8\pi n_i Z^2 r_e^2}{\gamma^2}\ln\left(\frac{b_{\max}}{b_{\min}}\right) = \frac{2k_p^2 r_e Z}{\gamma^2}\ln\left(\frac{\lambda_D}{R}\right), \tag{5.9}$$

where $n_i = n_e/Z$ is the ion density, Z is the charge state and $r_{\rm e}$ is the classical electron radius. The minimum and maximum impact parameters are estimated with $b_{\min} = R$ and $b_{\max} = \lambda_D$ where $R \approx 1.4\ A^{1/3}$ fm is the effective Coulombic radius, A is the atomic mass number and λ_D is the Debye length defined in Eq. (2.18) [93]. For an estimate, the weighted average of atomic number was used to calculate the effective Coulombic radius $R \approx 2.9$ fm and $Z \sim 100$. Assuming $n_{\rm e} \approx 4.35 \times 10^{23}\,\mathrm{cm}^{-3}$ and $T_{\rm e} \sim 100$ eV, the Debye length is $\lambda_D = 0.11$ nm. For a 200 MeV e-beam, $\gamma = 400$, the change in the rms divergence is $\mathrm{d}\langle\theta^2\rangle/\mathrm{d}z = 5.5 \times 10^{-6}\,\mathrm{mm}^{-1}$. For an e-beam travelling through 100 nm of the plasma, the change in the rms divergence is $\theta_{\rm rms} \sim 25$ μmrad. If the plasma were 35 μm thick as in the un-ionized tape, the scattering angle is $\theta_{\rm rms} \sim 439$ μmrad. This scattering will increase the emittance of the e-beam. For a 511 MeV ($\gamma = 1{,}000$) e-beam with $\varepsilon_N = 0.1$ mm-mrad, the 100 nm and 35 μm thick plasma placed 10 mm downstream of the 1st module increases the emittance by a factor of 3 and 40, respectively.

In the staging experiment, the plasma thickness will be a few skin-depths, closer to 100 nm. These scattering effects are particularly severe since e-beam energy in the current experiment is only a few hundred MeV. The emittance growth is mitigated for higher energy or smaller divergence e-beams due to the smaller scattering angle. In the case of a staged LPA that can reach ~10 GeV in the 1st module as discussed in references [27, 28], the scattering angle is only $\theta_{\rm rms} \sim 9$ μmrad for a 35 μm thick plasma. Furthermore, thinner foil of a few nano-meter thickness or a foil with a hole for the e-beam could be used to reduce the scattering.

5.5 Summary and Conclusions

In this chapter, the plasma mirror was introduced as a way to reduce the coupling distance and maintain the high average accelerating gradient in staged LPAs. Experimental results of the plasma mirror using VHS tape were discussed. The maximum reflectivity of ~80 % and Strehl ratio of 0.7, as good as input pulses, were observed at fluences of 300–700 J/cm^2 on the tape. To improve the reflected laser pointing stability, a monitoring diagnostic was implemented to assist in confirming real-time surface quality. As a result, a plasma mirror with sufficient reflective quality for the staging experiment was demonstrated.

The effect of a plasma mirror on the e-beam was discussed. The dynamical coupling of plasma wave and e-beam is expected to be small due to the much lower electron density in the e-beam than in the plasma mirror. The electron-ion scattering of e-beam in a polyester tape and a sheet of plasma is expected to increase the divergence of e-beams by a few hundred microradians in the staging experiment. This scattering will be significantly reduced for a higher energy or a narrower divergence e-beam as in future staged LPAs.

Chapter 6
Acceleration Module

6.1 Introduction

In this chapter, experiments and analyses related to efficient and controlled excitation of wakefields below the threshold of e-beam production in the 2nd module are presented. Properties of accelerating e-beams depend on amplitudes and shapes of plasma waves, which in turn are a function of drive laser and plasma channel properties. While large accelerating gradients are easily sustained in an LPA, precise knowledge of wakefield profiles, laser evolution and the longitudinal plasma channel profile are examples of physics that are yet to be understood, especially for stages below the trapping threshold as needed for staged LPAs.

Section 6.2 discusses a study of higher order modal contents in TREX laser pulses to evaluate their impact on wakefield profiles. Conventionally, the wake excitation is analyzed with Gaussian pulses. However, high power laser pulses often contain higher order aberrations. Wakefield profiles excited by a Gaussian pulse, as well as higher order modes, are compared with simulations.

Section 6.3 presents experiments on characterizing the plasma channel. Accurate knowledge of the matched spot size, r_{m}, and channel profile are critical in understanding the guided laser propagation and excited wakefield. However, the dynamic nature of plasmas makes characterization of the channel challenging. An improved experimental technique using laser centroid oscillation to characterize the channel is presented.

Section 6.4 discusses analyses of optical spectra and simulations to diagnose wakefields. Wakefield profiles and amplitudes have been experimentally measured using multiple shots and complex interferometric setups [94–96]. However, easier methods to gain insights into the excited wakefield and consequently the electron energy gain will be crucial for staged LPAs. One such method, a single shot optical spectral based diagnostic of the wakefield is presented.

Section 6.5 discusses an issue regarding splitting a high intensity laser pulse to drive multiple stages. In this staging experiment, the laser pulse is split after the full amplification and compression in the CPA system. This configuration results

S. Shiraishi, *Investigation of Staged Laser-Plasma Acceleration*,
Springer Theses, DOI: 10.1007/978-3-319-08569-2_6

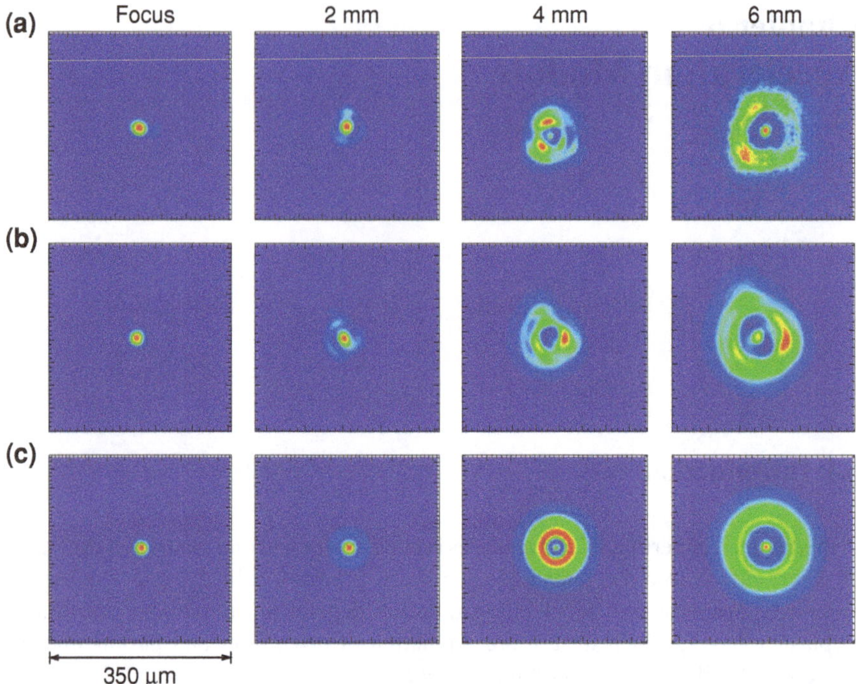

Fig. 6.1 Transverse intensity profiles of TREX pulses at focus, 2, 4, and 6 mm from the focus. **a** Measured intensity profiles using a CCD camera. **b** ZEMAX simulated intensity profiles based on measured wavefront and intensity using a wavefront sensor. **c** ZEMAX simulated intensity profiles with simplified wavefront and uniform intensity profile

in a high laser intensity at the splitting optic, which therefore modulates the pulse propagating through the optic. The basic physics of laser-matter interaction that cause the modulation, and the measured modulation in laser pulse 2, are presented. Ways to mitigate the modulation in future experiments are discussed.

Section 6.6 presents experimental results on wake excitation in the staging setup. Laser pulse reflected off the plasma mirror, propagated through the 2nd module, and excited wakefields. Based on the study described in Sect. 6.4, measured optical spectra are compared with simulations to gain insights into the accelerating field gradient. The results presented in this chapter are summarized in Sect. 6.7 along with a discussion of possible improvements for the wakefield excitation in the 2nd module.

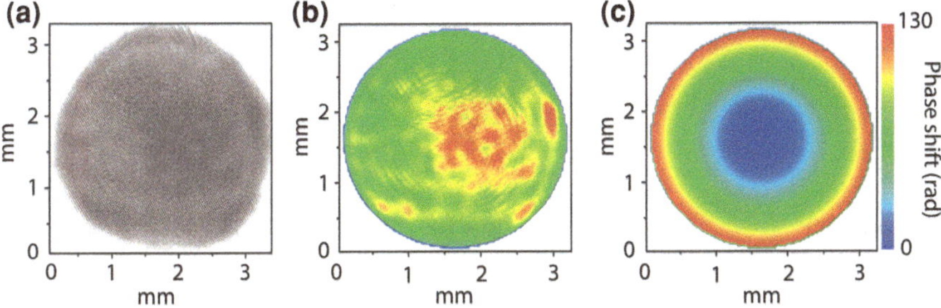

Fig. 6.2 **a** An interferogram of a TREX laser pulse measured using a SID-4 lateral shearing interferometer [98]. **b** Retrieved intensity and **c** wavefront maps. The interferogram was taken 80 mm upstream of the vacuum focus

6.2 Laser Profile Characterization

6.2.1 Wavefront Measurement

Wakefield shapes and focusing fields can depend on the spatial profile of drive lasers. In the linear to quasi-linear regime ($a_0 \lesssim 1$), the electric fields of the wake are proportional to the intensity profile. The longitudinal field is $E_z \sim \partial a^2/\partial z$ and the transverse field is $E_\perp \sim \nabla_\perp a^2$. This implies that the focusing force can be manipulated by tailoring the transverse laser profile. In Sect. 2.4, it was introduced that a large transverse field focuses the e-beam tightly and can cause electron energy loss through a large betatron oscillation in an LPA. Mitigation of the focusing field using a higher order mode was investigated theoretically and numerically by Cormier-Michel et al. [41]. The idea is to flatten the laser profile which leads to wakefield excitation of a weaker transverse field. Conversely, the transverse field can be steeper than expected from a Gaussian profile if the actual drive laser profile contains sharper intensity gradients. In this analysis, we investigate the effect of higher order modes on wake excitation in experiments.

High power laser pulses often contain higher order aberrations. Figure 6.1a shows a CCD camera based measurement of the laser intensity profiles at a focus, 2, 4, and 6 mm from the focus. Based on the spot size at focus, $r_0 = 18\,\mu\text{m}$, these TREX pulse can be modeled with Gaussian pulses with the Rayleigh length, $z_R = 1.3\,\text{mm}$. However, the deviation from a Gaussian profile is apparent around three z_R (4 mm). Wakefields excited by these non-Gaussian pulses were studied using simulations.

These laser profiles were simulated using ZEMAX optical simulation software based on experimentally measured wavefront and intensity [97]. An interferogram was measured ~80 mm upstream of the laser focus using a SID-4 lateral shearing interferometer [98]. This wavefront sensor splits the incoming pulse into four identical pulses and uses spectral analyses to retrieve a wavefront map from the interferogram. The interferogram, retrieved wavefront and intensity are shown in Fig. 6.2. The vacuum propagation of the wavefront and intensity was simulated using ZEMAX

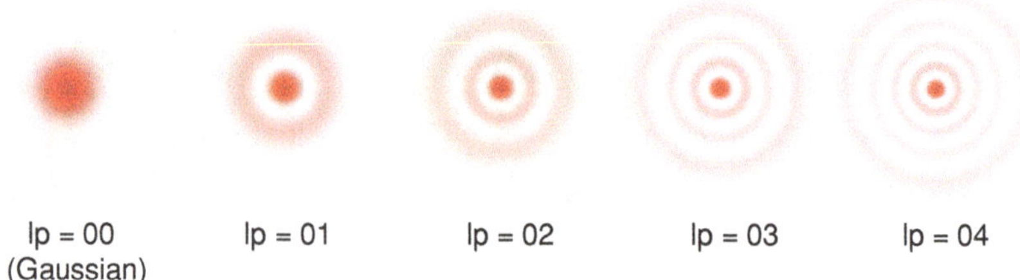

Fig. 6.3 Examples of axisymmetric Laguerre-Gaussian modes. The fundamental mode is the Gaussian distribution

and profiles are shown in Fig. 6.1b. The CCD measured (a) and simulated (b) intensity profiles agree well, indicating the accuracy of the phase measurement.

Non-axisymmetric aberrations were then removed in ZEMAX to symmetrize the pulse profile as required in the particle-in-cell (PIC) simulations for LPAs, and are shown in Fig. 6.1c. Each image was normalized to the peak intensity in the image. To achieve the cylindrical symmetry, the measured wavefront map was projected onto Zernike polynomials which are orthonormal polynomials defined on a unit disk. Zernike polynomials are often used to characterize a laser wavefront. In this analysis, polynomials of up to 45 coefficients were used and all non-axisymmetric coefficients were removed. The intensity was assumed to be a uniform distribution at the location of the wavefront measurement. These symmetrized profiles were closer to the experimentally measured profiles than a Gaussian distribution and allowed numerical investigation of wake excitations in LPAs.

6.2.2 Characterization with Laguerre-Gaussian Pulses

The symmetrized profiles shown in Fig. 6.1c were then characterized with up to fourth-order Laguerre-Gaussian (LG) modes to allow comparisons with theory. It was introduced in Chap. 2 that the Laguerre-Gaussian laser pulse is a solution to the paraxial wave equation if axial symmetry is assumed [32]. They are expressed in cylindrical coordinates using generalized Laguerre polynomials. Example modes are shown in Fig. 6.3. Electric fields for LG modes are expressed as in

$$u_{lp} = \frac{C_{lp}^{LG}}{r(z)} \left(\frac{\rho\sqrt{2}}{r(z)} \right)^{|l|} \exp\left(-\frac{\rho^2}{r^2(z)} \right) L_p^{|l|} \left(\frac{2\rho^2}{r^2(z)} \right) \exp\left(ik\frac{\rho^2}{2R_c(z)} \right) \\ \times \exp(il\phi) \exp\left[-i(2p + |l| + 1)\zeta(z) \right], \quad (6.1)$$

where $p \geq 0$ is the radial and l is the azimuthal index [99]. The C_{lp}^{LG} is the coefficient for each term and $L_p^{|l|}(x)$ is the generalized Laguerre polynomial. The radius of

Table 6.1 Generalized Laguerre polynomials up to 4th order

p	$L_p^0(x)$
0	1
1	$-x+1$
2	$(x^2-4x+2)/2$
3	$(-x^3+9x^2-18x+6)/6$
4	$(x^4-16x^3+72x^2-96x+24)/24$

Table 6.2 Coefficients for the LG decomposition of the TREX laser pulse

Parameter	r_0	C_{00}	C_{01}	C_{02}	C_{03}	C_{04}	ϕ_0	ϕ_1	ϕ_2	ϕ_3	ϕ_4
Fitted value	32.3	1	0.95	0.81	0.69	0.44	0	5.9	0.0	0.0	3.3

curvature $R_c(z)$ and longitudinal phase delay $\zeta(z)$ are as defined in Sect. 2.3. The intensity distribution assuming axisymmetry ($l=0$) up to 4th order LG modes is,

$$I(\rho,z)=\frac{1}{r(z)^2}\exp\left(-\frac{2\rho^2}{r(z)^2}\right)\left[\sum_{p=0}^{4}|C_{0p}|\exp(i\phi_p)L_0^p\left(\frac{2\rho^2}{r(z)^2}\right)\exp\left[-i\,(2p+1)\,\zeta(z)\right]\right]$$
$$\times\left[\sum_{p=0}^{4}|C_{0p}|\exp(-i\phi_p)L_0^p\left(\frac{2\rho^2}{r(z)^2}\right)\exp\left[+i\,(2p+1)\,\zeta(z)\right]\right], \qquad (6.2)$$

where $L_p^0(x)$ are listed in Table 6.1. The complex coefficients C_{0p}^{LG} are expressed as $|C_{0p}|\exp(i\phi_p)$.

The simulated profiles were then fitted with Eq. (6.2) to obtain LG coefficients that best described the pulse profile over more than 5 z_R (7 mm). Transverse lineouts of the intensity profiles at 0.5 mm increments up to 7 mm away from the focus were simultaneously fitted. Figure 6.4 shows the fits at six different locations in z, and the resulting LG coefficients are listed in Table 6.2. Figure 6.4 shows good agreement between the simulated profiles (black curve) and the fitted profiles (blue curve). For a given laser energy, the Gaussian pulse had a higher peak intensity than the LG pulse, and the difference was equivalent to the measured Strehl ratio at focus, 0.9. These LG modes modeled the axisymmetric, non-Gaussian features of the TREX pulses.

6.2.3 Wakefield Excitation by Gaussian and Laguerre-Gaussian Pulses

Amplitudes and shapes of wakefields excited by Gaussian and the LG pulses were compared through PIC simulations using INF&NO framework [36]. For the Gaussian pulse, the spot size $r_0 = 18\,\mu$m was assumed while the LG pulse was constructed from the fits described by Eq. (6.2) and Table 6.2. For both profiles, the initial temporal

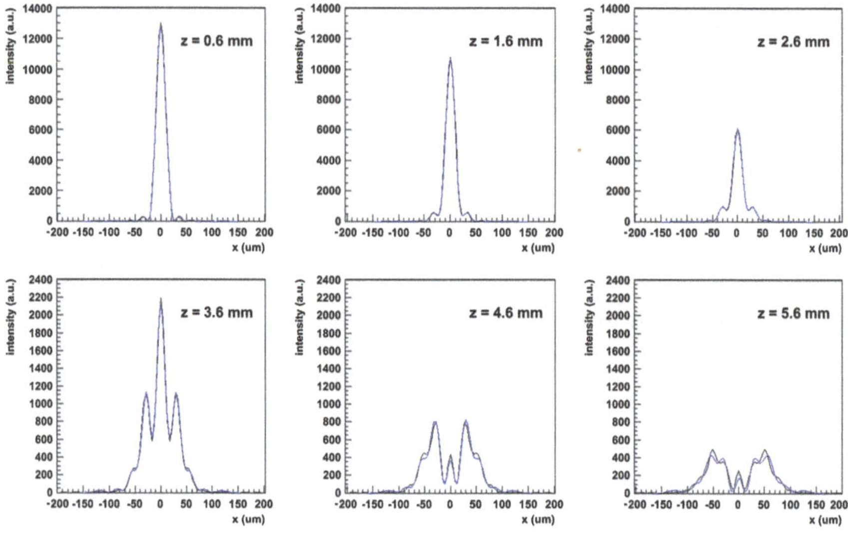

Fig. 6.4 Transverse laser profile fits at six different longitudinal positions where $z = 0$ is the laser focus. The *black lines* are from the ZEMAX simulated profiles and the *blue lines* are the fit results using Eq. (6.2)

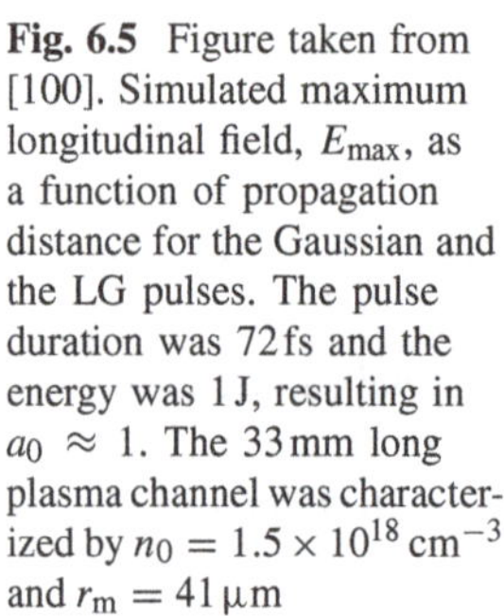

Fig. 6.5 Figure taken from [100]. Simulated maximum longitudinal field, $E_{\max}$, as a function of propagation distance for the Gaussian and the LG pulses. The pulse duration was 72 fs and the energy was 1 J, resulting in $a_0 \approx 1$. The 33 mm long plasma channel was characterized by $n_0 = 1.5 \times 10^{18}\,\text{cm}^{-3}$ and $r_m = 41\,\mu\text{m}$

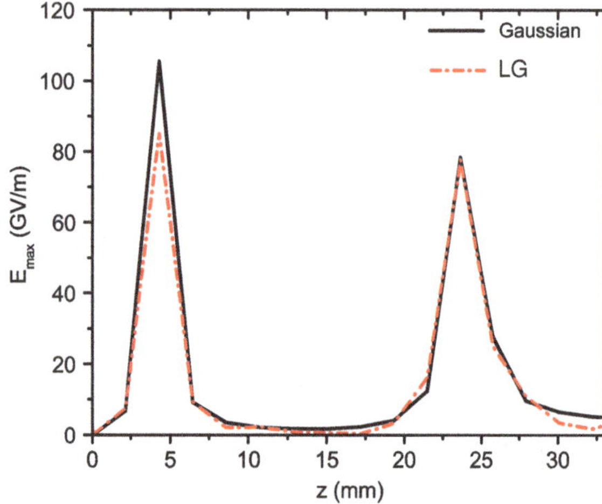

intensity profiles were Gaussian distributions of 72 fs at FWHM and the total energy in the pulses was 1 J. These conditions corresponded to $a_0 \approx 1.0$. The pulses were focused $Z_f = 4.8$ mm into the 33 mm long plasma channel which was characterized by parabolic plasma density profiles with $n_0 = 1.5 \times 10^{18}\,\text{cm}^{-3}$ and $r_m = 41\,\mu\text{m}$. These parameters are below the injection thresholds and were used in experiments.

The Gaussian and the LG pulses excited wakefields with similar maximum longitudinal electric field, $E_{\max}$, along the plasma channel as shown in Fig. 6.5.

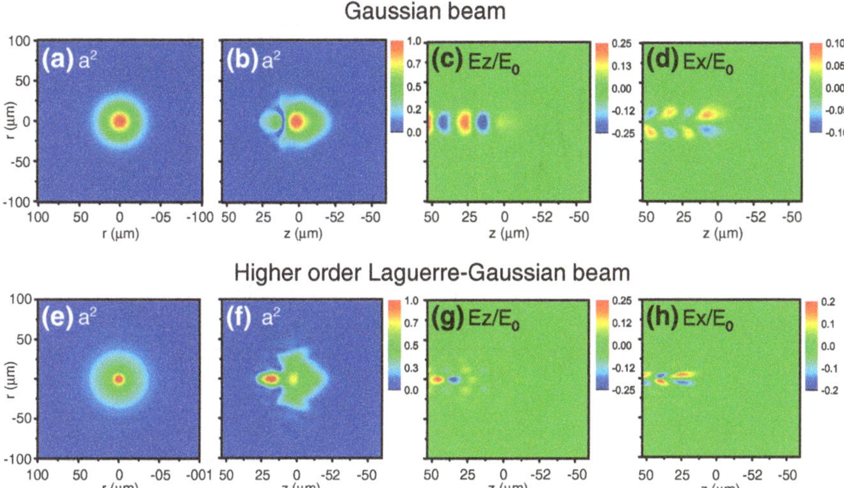

Fig. 6.6 Figure taken from [100]. Laser field and wakefield profiles for the Gaussian and LG pulses at $z = 26$ mm, approximately $2\,z_R$ from the second excitation peak. The top row shows the fields for the Gaussian pulse and the bottom row shows the fields for the LG pulse. The laser pulse propagates toward positive z. **a** and **e** are time integrated a^2 distribution, proportional to laser intensity. In the r–z coordinate, **b** and **f** are a^2, **c** and **g** are E_z/E_0, and **d** and **h** are E_x/E_0 where E_0 is 117 GV/m

The similar E_{max} suggests that the higher order modal contents do not significantly alter the a_0 evolution from that of the Gaussian pulse in the plasma channel for these parameters. It is important to note that the actual laser mode used in experiments contained non-asymmetric aberrations as shown in Fig. 6.1a. These aberrations could have reduced a_0 and caused E_{max} to evolve differently in the experiment than in these simulations.

While E_{max} evolutions were similar for both pulses, significantly different wakefield profiles were observed at some locations in the channel. Laser intensity and wakefields at $z = 26$ mm, approximately $2\,z_R$ from the second excitation peak, are shown in Fig. 6.6. The top row are the fields excited by the Gaussian pulse and the bottom row are those by the LG pulse. The laser pulse propagates towards $+z$. In the figure, (a) and (e) are longitudinally integrated a^2 which is proportional to the intensity. The laser a^2 [(b) and (f)], normalized longitudinal field E_z/E_0 [(c) and (g)] and normalized transverse field E_x/E_0 [(d) and (h)] are shown in the r–z coordinate. The electric fields are normalized to the non-relativistic, cold wavebreaking field $E_0 = 117$ GV/m. These figures show that the wakefields are transversely smaller for the LG pulse than the Gaussian pulse due to the narrower a^2 peak as shown in Fig. 6.6e compared to (a).

Longitudinal evolution also differs between the two pulses. The intensity peak is near the middle of the temporal profile for the Gaussian pulse (b) whereas for the LG pulse, the intensity peak is at the back of the laser pulse (f). This is probably due to mode slippage in the LG pulse. The different modal components of the LG pulse

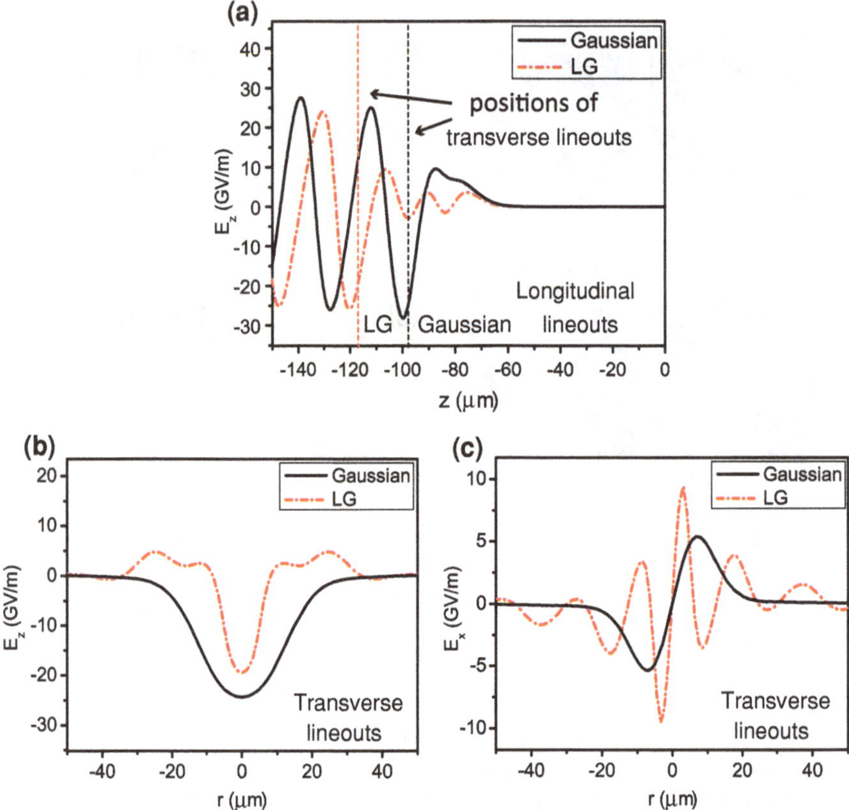

Fig. 6.7 The lineouts of longitudinal and transverse fields from Fig. 6.6. **a** The longitudinal lineouts are for $r = 0$. The transverse E_z lineouts **b** and E_x lineouts **c** for $z = -117\,\mu\text{m}$ ($k_{\text{p}}z = -27.2$) for the LG pulse and $z = -97\,\mu\text{m}$ ($k_{\text{p}}z = -22.75$) for the Gaussian pulse

propagate at different group velocities in the plasma [101]. Since the wakefields are excited by the intensity gradient, the difference in temporal intensity profiles can then result in different wake amplitudes. Detailed theory of wake excitation enhancement and experiments on enhanced electron injections by temporally skewed pulses are described in [102–104]. While E_{max} was similar for the two pulses considered here, the different longitudinal pulse evolutions due to the higher order modes can lead to different wake amplitudes.

Lineouts of the fields are shown in Fig. 6.7. Transverse lineouts for E_{z} and E_{x} are taken at $z = -117\,\mu\text{m}$ for the LG pulse and $z = -97\,\mu\text{m}$ for the Gaussian pulse, and are shown in Fig. 6.7b, c. The transverse extent of the E_{z} excited by the LG pulse is $\sim$1/3 of that of the Gaussian. Furthermore, the amplitude of E_{x} excited by the LG pulse is twice that by the Gaussian pulse. These differences imply that e-beam properties such as charge and divergence can be different for LPAs driven by a Gaussian pulse and an LG pulse.

For these laser and plasma parameters, this study suggests that E_{max} is similar for the Gaussian and the LG pulses. However, a study of phase velocities of wakefields and e-beam properties such as charge and divergence require consideration of higher order modal contents.

6.3 Plasma Channel Characterization

6.3.1 Plasma Channel Formation

A plasma channel in a capillary discharge is a dynamic waveguide that evolves in the nanosecond time scale. For experiments discussed in this thesis, capillaries were laser machined in sapphire plates at diameters of 200–300 μm. The length of the capillaries were 15 or 33 mm. Hydrogen gas (or mixed gas for injection study as discussed in Chap. 4) flowed through slots located ~2 mm from each end, and an electric discharge was fired prior to the arrival of the laser pulses to ionize the gas and form the plasma. Detailed theoretical and numerical investigation of the capillary discharge waveguides are found elsewhere [47, 59, 60]. The plasma evolution in the capillary during the electric discharge can be described in three stages [47]. During the initial ionization phase which lasts ~50 ns, the electron plasma temperature T_e is low and heat transfer to the wall of the capillary is slow. This results in a uniform plasma temperature and a radial electron density distribution n_e.

The second stage of channel formation begins near the full ionization of the plasma at $T_e \approx 2$ eV and takes place 50–100 ns after the initiation of the discharge. Ohmic heating (also referred to as resistive heating) causes a rapid rise in electron temperature. Since the capillary wall, made of sapphire, conducts heat, the temperature distribution is inhomogeneous: T_e is hot near the center of the channel and cold near the wall. Based on the ideal gas law, this temperature gradient is the pressure gradient to redistribute electrons. The result is a parabolic density profile with a lower density on axis and a higher density near the wall which can be described by the on-axis electron density n_0 and the matched spot size r_m as shown in Eq. (2.10).

The third stage begins ~100 ns after the electric discharge, and the plasma channel settles to a stable quasi-equilibrium state in the radial direction. This is the regime in which LPA experiments are performed. Detailed experimental investigations of the plasma channel are found elsewhere [105, 106]. For a typical discharge pulse of 250 A peak and 400 ns wide at FWHM, this stable quasi-equilibrium state lasts for ~300 ns. The axial T_e reaches its maximum at the time of maximum electric current and gradually decreases with time. For a current with 250 A with capillary diameter of 250 μm, $T_e \sim 7$ eV at the peak based on a simulation [47]. When T_e cools down to 1–2 eV, the recombination of the hydrogen begins and the parabolic density distribution disappears.

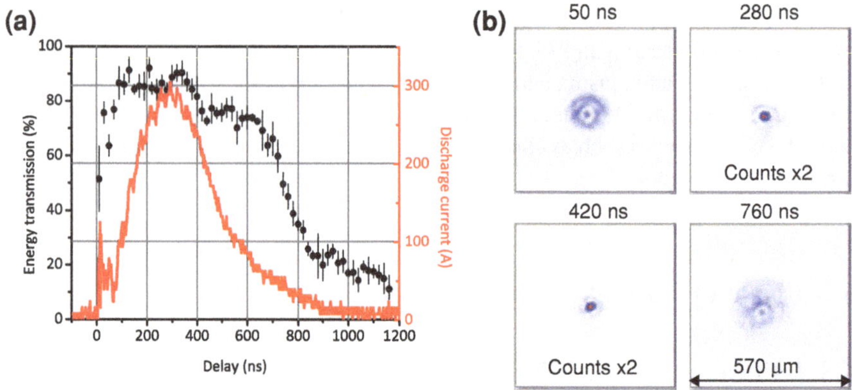

Fig. 6.8 **a** Measured laser energy transmission as a function of delay (*black*) and the discharge current (*red*). **b** Output laser modes at the exit of the capillary for different delays. Delays between 100 and 600 ns demonstrated good guiding with high transmission and good output mode

The formation and deformation of the waveguide can be indirectly observed by monitoring the laser properties propagated through the capillary [107]. Figure 6.8a shows the discharge current and measured laser energy transmission as a function of the delay with respect to the initiation of the discharge (Delay $= 0$ ns). For this experiment, low power pulses of $a_0 \sim 0.01$ were guided through a plasma channel of $n_0 \sim 1.5 \times 10^{18}\,\text{cm}^{-3}$ and $r_m \approx 41\,\mu\text{m}$ [59]. Before the formation of the parabolic channel (<100 ns), the energy transmission fluctuates significantly and the output laser profile is poor as shown in Fig. 6.8b 50 ns. When the parabolic channel is formed and guides the laser (100–450 ns), the laser propagates through the capillary without significant energy loss indicated by the transmission of ~90 %. The output mode in Fig. 6.8b 280 ns and 420 ns shows clear intensity peaks. For this particular scan, there is a plateau in the energy transmission of ~75 % from 450 to 600 ns. This may be due to the high peak discharge current (~300 A) that delays the cooling of the electrons and consequently the recombination process. After >650 ns, the transmission decreases rapidly and the output profile degrades as shown in Fig. 6.8b 760 ns. The observation of laser energy transmission and output modes are one of the ways to optimize laser guiding in an evolving plasma channel [107].

6.3.2 Laser Centroid Oscillation

The centroid oscillation of the laser exiting the plasma channel can be used to measure r_m, plasma channel profile, and to align the laser with respect to the channel. The experimental results presented here were published in Gonsalves et al. [105]. Experimental and numerical studies have given scaling laws for on-axis plasma density n_0 and matched spot size r_m as functions of the capillary radius r_c and the initial

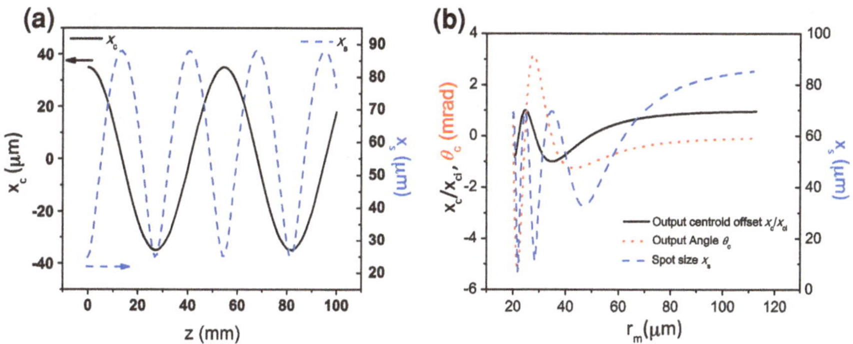

Fig. 6.9 Figures from [105]. **a** Laser centroid offset and laser spot size in x-direction, x_s as a function of propagation distance. Matched spot size $r_m = 47\,\mu m$, $x_{ci} = 35\,\mu m$, and $r_i = 25\,\mu m$ was used for the calculation. **b** Laser parameters at the output of a 15 mm plasma channel as a function of r_m. Normalized laser centroid offset x_c/x_{ci} (*solid line*), angle (*dotted line*) and spot size (*dashed line*) are shown for $x_{ci} = 10\,\mu m$, $\theta_i = 0$ and $r_i = 80\,\mu m$

hydrogen density $n^i_{H_2}$ [47, 59, 60]. Although these studies agreed on the scaling laws, the estimated uncertainty was as high as 20 %, mainly due to the uncertainty in phase retrieval from interferograms [60]. Since an understanding of the plasma channel is critical to understanding wake excitation, a more precise method was developed using laser centroid oscillations.

When the laser enters the plasma channel with a transverse offset x_{ci} and angle θ_i with respect to the channel axis, the laser centroid oscillates within the plasma and the spot size oscillates around the centroid. The evolution of the laser spot size in a parabolic channel is described by Eq. (2.14). The centroid oscillation is described as [105],

$$\left(\frac{\partial^2}{\partial \hat{z}^2} + \frac{\Delta n}{\Delta n_c}\right)\hat{x}_c = 0, \tag{6.3}$$

where $\hat{x}_c = x_c/r_m$ and Δn is the channel depth at a r_m. A solution to Eq. (6.3) is $x_c = x_i \cos[k_{\beta c}z - \phi]$, where $x_i = [x_{ci}^2 + \theta_i^2 k_{\beta c}^{-2}]^{1/2}$, initial phase $\phi = \arccos(x_{ci}/x_i)$ and $k_{\beta c} = Z_m^{-1}$ with $Z_m = \pi r_m^2/\lambda$. Figure 6.9a shows laser centroid offset and laser spot size in x-direction x_s as functions of the propagation distance for initial spot size $r_i = 25\,\mu m$, $r_m = 47\,\mu m$ and $x_{ci} = 35\,\mu m$. This dependence of r_m, x_{ci} and θ_i on the centroid oscillation allows for the measurement of r_m and the alignment of the laser with respect to the channel ($x_{ci} = 0$ and $\theta_i = 0$) using the oscillation.

The matched spot size r_m is measured by observing the laser centroid shift as a function of discharge delay. Figure 6.9b shows output laser parameters as a function of r_m. The output centroid offset normalized by input offset x_c/x_{ci} (solid line), the angle θ_c (dotted red line), and the spot size x_s (dashed blue line) at the exit of a 15 mm plasma channel are shown. The values are calculated from Eq. (6.3) and Eq. (2.13) for $\theta_i = 0$, $x_{ci} = 10\,\mu m$ and $r_i = 80\,\mu m$. For a measured x_{ci} or x_s, there are multiple solutions to the r_m. In experiments, this ambiguity is resolved by

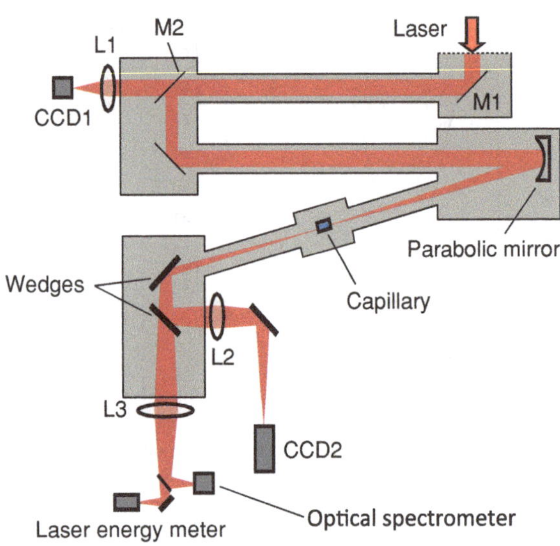

Fig. 6.10 Figures from [105]. Experimental layout used for plasma channel characterization. TREX laser pulses are focused onto a capillary. Post-interaction pulses are attenuated by two glass wedges and their modes, energies and spectra are measured

scanning the discharge delay. The assumption is that although the plasma channel acts as a waveguide, the channel depth continuously decreases and r_m increases on the nanosecond timescale after the peak of the discharge current [105]. Consequently, scanning the delay from a large delay towards the current peak effectively scans the r_m from infinity to a smaller r_m. For a parabolic plasma channel, r_m as a function of the centroid offset is,

$$r_m = \sqrt{\frac{2z}{k|\cos^{-1}(dx_c/dx_{ci}) + 2\pi j|}}, \tag{6.4}$$

where the integer j can be determined by measuring the response of the laser to discharge timing. By measuring the output centroid offset as a function of x_{ci} and the discharge timing, r_m is deduced.

The centroid oscillation is also used to measure the channel profile. For a hydrogen-filled capillary discharge waveguide, the density profile for $r < r_m$ has been predicted to be parabolic, but rises faster than quadratically near the capillary wall [47,108]. For a fixed channel length, the centroid shift $x_{shift} = x_c - x_{ci}$, is a linear function of x_{ci} when the channel is parabolic as can be seen in the solution to Eq. (6.3). When the channel profile is quadratic, $n(r) = n_0 + \Delta n r^4/r_m^4$, the x_{shift} dependence on x_{ci} becomes nonlinear since the laser is deflected strongly. By measuring the dependence of on x_{ci}, the plasma channel shape is inferred.

6.3.3 Experimental Configuration

Experiments were performed to demonstrate these concepts and the setup is shown in Fig. 6.10. Low energy pulses (<5 mJ) from TREX laser systems were focused onto the capillary by a 2 m focal length off-axis parabolic mirror (OAP). An aperture 2 cm in diameter was used to increase the effective f-number of the focusing system, resulting in a laser spot size of $r_{\mathrm{i}} = 70\,\mu\mathrm{m}$ at the capillary entrance. This large spot size implied that $z_{\mathrm{R}} = 19\,\mathrm{mm}$, allowing for the unguided pulse to pass through the 15 mm long, 300 μm diameter capillary with minimal wall interaction. Therefore, the laser centroid location at the output of the capillary when there is no plasma channel was the same as when the capillary was removed. The peak intensity on target was $5 \times 10^{13}\,\mathrm{W/cm^2}$ for a pulse length of 120 fs. This low intensity and low power condition simplified the laser-plasma interaction by avoiding ionization of hydrogen and self-focusing. The plasma channel was formed with $n_0 \simeq 1.4 \times 10^{18}\,\mathrm{cm^{-3}}$ using the scaling law given in [60]. Laser exiting from the capillary was attenuated by reflection off two optically flat glass wedges and refocused by a lens of focal length 500 mm and diameter 100 mm. This imaged the laser at the exit of the capillary onto a 12-bit CCD camera to measure the centroid position x_c and spot size x_{s}.

The capillary was first aligned by observing the laser output mode while varying the capillary transverse position and angle. For optimum alignment, translation of the capillary position by an equal amount in either direction resulted in an equal change in the portion of laser energy transmitted through the waveguide, and of the shift in the laser centroid x_{shift}. The x_{shift} was experimentally defined as the difference in the centroid positions between the pulses at the output plane of the capillary without the plasma channel and the pulses propagated through the waveguide. The alignment was confirmed when the x_{shift} was constantly zero as a function of discharge delay, suggesting $x_{\mathrm{ci}} = 0$ and $\theta_{\mathrm{i}} = 0$.

6.3.4 Experimental Results and Analysis

The r_{m} was measured based on the centroid offset x_{shift} and on the spot size x_{s} independently and both results agreed with simulations. The laser centroid was intentionally offset $x_{\mathrm{ci}} = 30\,\mu\mathrm{m}$ and the x_{shift} was measured. The measured x_{shift} at the capillary exit as a function of discharge timing is shown as blue triangles in Fig. 6.11a. The discharge current with ~350 A peak and ~400 ns long at FWHM is shown as the green curve. The x_{shift} increases to $\sim -25\,\mu\mathrm{m}$ during the first 200 ns and stabilizes between the delay of 200 and 600 ns. For delays after this, when the discharge current drops below 100 A, the x_{shift} begins to return to zero from hydrogen recombination. The r_{m} was measured by scanning backward in delay which corresponds to scanning from infinite to finite r_{m} in Fig. 6.9b. The r_{m} can be uniquely determined with the assumption $\theta_{\mathrm{i}} = 0$ using Eq. (6.4) with $\mathrm{d}x_c/\mathrm{d}x_{\mathrm{ci}} = x_c/x_{\mathrm{ci}}$ and $j = 0$. The calculated r_{m} based on x_{shift} is shown as black squares in Fig. 6.11a. For delays where

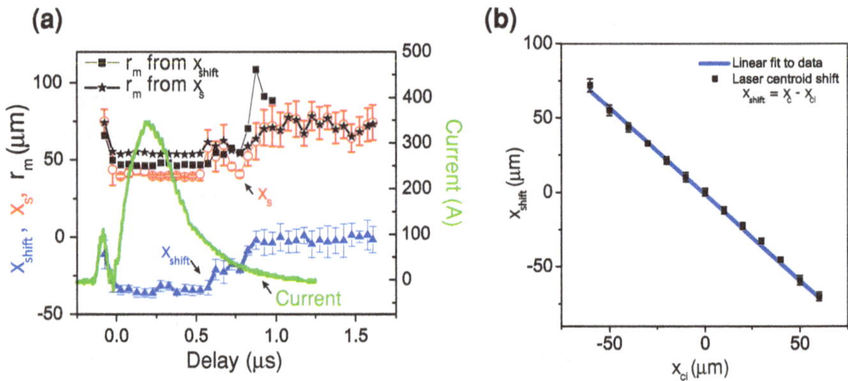

Fig. 6.11 Figures from [105]. **a** The centroid shift x_{shift} (*blue triangles*), spot size x_s (*red circles*) and r_m (*black squares and starts*) as a function of delay between the discharge and the laser arrival. The discharge current is also shown with a *green curve*. **b** x_{shift} as a function of x_{ci}, indicating a parabolic channel profile for $r < r_m$

the current is above 100 A, the average matched spot size was $r_m = 46.9 \pm 0.8\,\mu\text{m}$, which agrees with 46.5 μm given by the scaling derived from simulations [59]. The output spot size x_s is shown as red circles in Fig. 6.11a. The r_m was calculated using Eq. (2.14) and shows similar results as those from x_{shift}. The difference between the two sets of measurements are in part due to the laser mode not being a perfect Gaussian distribution. Although Eq. (2.14) is valid for all higher order modes, the recovery of r_m from the fitted spot size is not straightforward when the laser contains higher order modes. While x_{shift} was a more precise measure, both methods showed good agreement with the scaling law given by the simulation.

The laser centroid oscillation indicated a parabolic plasma channel profile over the extent of the matched spot. The discharge delay was fixed at 225 ns and x_{ci} was varied by transversely moving the capillary to measure its dependence on x_{shift}. Measured x_{shift} showed a linear dependence on x_{ci} as shown in Fig. 6.11b. It was introduced earlier in this section that this linear dependence implied a parabolic plasma channel profile for $r < r_m$.

The alignment technique using the centroid oscillation is an improvement over methods relying on laser energy throughput or modal shapes. For a properly aligned channel, $\theta_i = 0$ and $x_{ci} = 0$, the x_{shift} is 0 regardless of discharge delay (or r_m). By adjusting the capillary until no dependence of discharge delay on x_{shift} is observed, the channel can be aligned to the laser pulse. Alignment methods that rely on optimizing energy throughput or modal shape typically allow precision to within ~20 μm [105]. On the other hand, the accuracy of the centroid methods is limited by the accuracy of centroid determination. In this experiment, resolution was ~5 μm and limited by the magnification of the imaging system. This improvement in alignment is crucial to staged LPAs because of the stringent tolerance on the e-beam alignment with respect to wakefields, which is strongly influenced by the laser alignment with respect to the plasma channel.

This study measured r_m and the channel profile and were based on the physics of laser centroid oscillation in the plasma channel. The centroid oscillation also allowed for an improved alignment of the channel with respect to the laser.

6.4 Wakefield Diagnostic Based on Laser Spectra

6.4.1 Background on Wakefield Diagnostics

Measuring wakefield amplitudes in a channel guided LPA is difficult due to the complex dynamics of the laser pulse. Techniques such as frequency-domain interferometry and frequency-domain holography can measure wakefield structures. However, such experiments require probe pulses and multiple shots to characterize wakefields observed by e-beams for many Rayleigh lengths [94–96]. Optical spectrum shifts of a driving laser have also been used as wakefield diagnostics. The spectral red-shifts are directly related to laser energy depletion in the plasma. Energy transferred from the laser to the plasma is converted to energy in longitudinal and transverse wakefields. Murphy et al. detected a large amplitude wakefield through "photon acceleration [109]." This technique requires the laser pulse to be longer than optimal for wake excitation. For the regime discussed in this section, the laser pulse was too short to experience this effect. More recently, optical spectra have been used to measure excited wakefields in gas-filled capillary tubes. This analysis reported good agreement between experiment and theory for the case of wake excitation in a linear regime where the laser pulse was guided in the channel without significant transverse evolution [110].

In this analysis, spectral shifts are studied to diagnose wakefields in a plasma channel with strong laser evolution resulting in linear and nonlinear wakes. Without transverse laser evolution, the ratio of energy going into the longitudinal and transverse fields is nearly constant. However in many experiments, the transverse evolution of the laser can be significant, resulting in a varying partition of laser energy into the longitudinal and transverse fields as a function of propagation distance. Therefore, understanding both the efficiency of the total laser energy transfer into the plasma, and the energy partition between longitudinal and transverse fields are important for estimating the accelerating fields in an LPA. Our analysis identifies some of the critical experimental parameters that influence the efficiency of the laser energy transfer into the plasma. Specifically, the influence of input laser energy, plasma density, temporal and spatial laser pulse shape, and laser coupling condition defined by laser focus position and longitudinal plasma density profile are studied experimentally and numerically. In addition to the expected increase in the energy transfer efficiency with laser intensity and plasma density, the efficiency is found to be particularly sensitive to the temporal pulse shape and the laser coupling condition. The experimental conditions for these parameters are modeled and included in particle-in-cell (PIC) simulations, resulting in an agreement of simulated redshift with experimental data within uncertainties. The simulations are then used to estimate the wake amplitudes in the experiments.

Section 6.4.2 will introduce the basic physics of spectral redshift. Section 6.4.3 will present the experimental configuration. Section 6.4.4 discusses an example of simulations performed with experimental parameters. In Sect. 6.4.5, the analysis of experimental data and simulation results are presented. Section 6.4.6 summarizes the results and discusses the implications of the analysis. These experimental results were published in Shiraishi et al. [100].

6.4.2 Spectral Redshift as a Measure of Wake Excitation

The redshift of the driving laser pulse is correlated with the overall laser energy transferred to the plasma for wake excitation. In one-dimensional theory, the evolution of the laser energy, ε, of a short laser pulse propagating in an underdense plasma is related to the amplitude of the wakefield by [56, 111]

$$\frac{\partial \varepsilon}{\partial \omega_{\mathrm{p}} t} = -\frac{k_{\mathrm{p}}^2}{k_0^2}\left(\frac{E_{\max}(z)}{E_0}\right)^2, \tag{6.5}$$

where ω_{p} and k_{p} are the plasma frequency and wavenumber, k_0 is the central wavenumber of the input laser, and $E_{\max}/E_0$ is the maximum accelerating field normalized by the cold nonrelativistic wave breaking field $E_0 = m_{\mathrm{e}} c\, \omega_{\mathrm{p}}/e$. To see the relation between the wavenumber shift and the wakefield amplitudes, we define the normalized laser energy as $\varepsilon \sim \int (k/k_0)^2 \left|\hat{a}(k)\right|^2 \mathrm{d}k$ and the wave action as $A \sim \int (k/k_0) \left|\hat{a}(k)\right|^2 \mathrm{d}k$, where $\hat{a}$ is the laser pulse envelope. The wave action A is an adiabatic invariant of the laser-plasma interaction resulting from the time-scale separation $k_{\mathrm{p}} \ll k_0$. We define the normalized average wavenumber as

$$\bar{k}/k_0 \equiv \frac{\int \left(\frac{k}{k_0}\right)^2 \left|\hat{a}(k)\right|^2 \mathrm{d}k}{\int \frac{k}{k_0} \left|\hat{a}(k)\right|^2 \mathrm{d}k} \simeq \varepsilon/A. \tag{6.6}$$

Since $k_{\mathrm{p}} \sim 0.2\,\mu\mathrm{m}^{-1}$ and $k_0 \sim 7.9\,\mu\mathrm{m}^{-1}$, the action A is approximately constant in our experiments. Equation (6.6) shows that laser energy depletion is proportional to the spectral redshift. This equation is valid in three-dimensions. In this analysis, the amount of redshift is calculated as

$$\Delta R = 1 - \bar{k}_{L_{\mathrm{ch}}}/k_0, \tag{6.7}$$

where $\bar{k}_{L_{\rm ch}}/k_0$ is calculated using Eq. (6.6) at the exit of the channel. Equations (6.5) and (6.6) relate ΔR to the wake amplitude,

$$\Delta R \propto k_{\rm p}^3 L_{\rm ch} \left\langle \frac{E_{\rm max}^2}{E_0^2} \right\rangle, \tag{6.8}$$

where $L_{\rm ch}$ is the length of the channel and $\langle f \rangle = 1/L_{\rm ch} \int_0^{L_{\rm ch}} f(z) {\rm d}z$ is the average of the function $f(z)$ over the length of the plasma channel. Since spectral shifts are less than 10 % for the cases considered in this article, ΔR roughly equals $\Delta\lambda/\lambda_0$. Furthermore, $E_{\rm max}/E_0 \simeq a_0^2(1+a_0^2)^{-1/2}$ in the one-dimensional limit [5]. Using $k_{\rm p}^3 \propto n_0^{3/2}$, this implies $\Delta R \sim a_0^4 n_0^{3/2}$ in the linear regime and $\Delta R \sim a_0^2 n_0^{3/2}$ in the nonlinear regime. This relation will be used to guide the analysis. The spot size evolution changes the energy partition between the transverse and longitudinal fields. In this case, the accelerating field cannot be inferred from ΔR alone. While the linear relationship between ΔR and $E_{\rm max}^2$ holds only if the spot size and wakefield amplitude are constant, the instantaneous change in spectral shift is always related to $E_{\rm max}^2$ regardless of the transverse evolution.

6.4.3 Experimental Configuration

The experimental configuration for measuring the optical spectra of driving laser pulses in the LPA is shown in Fig. 6.12. Pulses from the TREX laser system were focused using an off-axis parabolic mirror at f/22. Vacuum focal spot size was $r_0 =$ 18 μm and the Strehl ratio was 0.91. The Rayleigh length of the laser was $z_{\rm R} =$ 1.3 mm. The laser energy at the target ranged from 0.4 to 1.8 J. Temporal pulse shapes and durations were measured with a GRENOUILLE [61], and the pulse duration ($\Delta\tau$) ranged from 44 to 160 fs at FWHM. For the experiments discussed in this section, wake excitation was performed below the electron injection threshold.

A hydrogen-filled capillary discharge waveguide was used to form a plasma channel. A schematic of the capillary and estimated longitudinal density profiles are shown in Fig. 6.12b. The capillary was 250 μm diameter and 33 mm long. Hydrogen gas was inserted through slots located 2 mm from each end and an electric discharge was fired 200 ns before the arrival of the laser pulses. The peak of the discharge current was 260 A and the duration was 340 ns at FWHM. The laser arrived at the capillary ~50 ns before the peak of the current. The n_0 ranged from 0.7 to 2.1 × 10^{18} cm^{-3} based on the scaling law given by Broks et al. in Ref [59]. It is also important to consider longitudinal density gradient effects at the ends of the capillary [107]. This effect was investigated with simulations using two density ramps shown in Fig. 6.12b and will be discussed in Sect. 6.4.5.

Post-interaction laser pulses exiting the capillary waveguides were diagnosed using an optical spectrometer. The laser was attenuated after exiting the capillary by reflection off two spectrally flat glass wedges. The pulses were re-focused by a lens

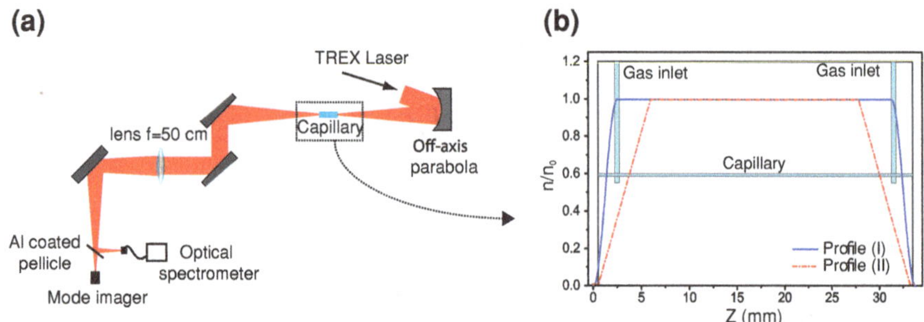

Fig. 6.12 Figures from [100]. **a** The experimental layout. The laser pulse was focused onto a capillary waveguide using a 2 m focal length off-axis parabolic mirror. The output laser pulse was attenuated with two uncoated wedges and weakly focused onto the optical spectrometer. **b** Schematic of the capillary (*grey rectangle*) and gas inlets. Normalized on-axis plasma density as a function of propagation distance for two different profiles (I) and (II). The effects of two profiles are investigated in Sect. 6.4.5

of focal length 500 mm and diameter of 100 mm for the post-interaction diagnosis. A fraction of the laser pulse was reflected off an aluminum coated pellicle at a laser pulse diameter of ~10 mm and focused onto a fiber based Hamamatsu Mini-spectrometer [112]. The wavelength dependent sensitivity of the detector and the wavelength dependence of the transport beamline were calibrated using a light source with a known spectrum. After the calibration, the intensity of the laser pulse as a function of wavelength was obtained. Post-interaction optical spectra were measured for each laser shot, and the dependence of the redshift on LPA parameters—a_0, n_0, laser profiles, and focal location—were studied.

Data used in this analysis were categorized in four sets and are listed in Table 6.3. Sets 1 and 2 were analyzed to study the amount of optical redshift as a function of laser intensity and plasma density. The pulse was positively chirped and the duration was 72 fs at FWHM. Laser pulse energy varied from 0.4 to 1.8 J which corresponds to $a_0 = 0.8$–1.3 assuming a Gaussian pulse and accounting for the Strehl ratio. The n_0, for delays after a stable plasma channel was formed, varied from 0.7 to $2.1 \times 10^{18}\,\text{cm}^{-3}$. These data sets allowed us to study the amount of redshift as a function of a_0 and n_0. The third data set was analyzed to study the effect of temporal pulse shapes. For this experiment, the laser pulse energy was fixed at 1.8 J and $n_0 = 1.5 \times 10^{18}\,\text{cm}^{-3}$. The pulse duration and temporal shape were varied by changing the distance between the compressor gratings. The fourth set of data allowed us to investigate the effect of laser focus position Z_f with respect to the longitudinal plasma density ramps.

Table 6.3 Parameters for the experimental data set

Data set	a_0	n_0 (cm^{-3})	$\Delta\tau_{\text{FWHM}}$ (fs)	Z_f (mm)
1	0.8–1.3	1.5×10^{18}	72	4.8
2	1.2	$0.7\text{–}2.1 \times 10^{18}$	72	4.8
3	0.9–1.7	1.5×10^{18}	44–160	4.8
4	1.2	1.5×10^{18}	72	0–4.8

The laser energy was varied for set 1. The hydrogen gas pressure was varied for set 2. The separation of the compressor gratings was varied for set 3. The capillary position with respect to the vacuum laser focus was varied for set 4

6.4.4 Simulations

The experiments discussed in this article were performed with mismatched guiding, resulting in a strong laser evolution. The guiding introduced significant transverse spot size oscillation, and the focusing of the laser resulted in $a_0 \gtrsim 1$. Simulations using the INF&RNO framework were used to investigate the effects of laser evolution with the experimental conditions [36]. Laser-plasma interactions are described using the ponderomotive-force approximation, and for the set of simulations used in this analysis, a quasi-static approximation was applied [5]. An example of simulated accelerating fields and optical spectra using experimental parameters are shown in Fig. 6.13. The laser pulse was focused 4.8 mm into the channel with vacuum focus of $r_0 = 18\,\mu$m and $a_0 = 1.2$. A Gaussian distribution was assumed for the transverse laser profile, and the a_0 was adjusted to account for the measured Strehl ratio. Measured temporal pulse shape and spectral phase were used in the simulation and the duration was 75 fs at FWHM. The plasma channel was described by $r_m = 41\,\mu$m, $n_0 = 1.5 \times 10^{18}\,\text{cm}^{-3}$, and 33 mm in length. The $E_{\max}$, $\bar{k}$ and ε are shown as functions of propagation distance in Fig. 6.13a. Since the guiding was mismatched, the laser spot size and the peak intensity oscillated. There were two regions where the laser was strongly focused to excite a large plasma wave. The ε curve showed large energy transfer from the laser to plasma at locations of large wakefield excitations. Moreover, the $\bar{k}$ and ε curves closely tracked each other as anticipated from the theory. Figure 6.13b shows optical spectra at three different locations in the simulation. The blue curve is the input spectrum, centered around 800 nm. After the large wake excitations, the spectra are shown as the green and red curves. This simulation illustrates the excitation of large wakefields at two regions ($z \sim 5$ and 25 mm) in the channel correlating strongly with the laser redshifting.

The degree of energy partition between the transverse and longitudinal fields is a function of a_0. Figure 6.13c shows the ratios of energy in the fields defined as

$$\frac{U_\perp}{U_\parallel} = \frac{\int(|E_r|^2 + |B_\theta|^2)\mathrm{d}V}{\int |E_z|^2 \mathrm{d}V} \tag{6.9}$$

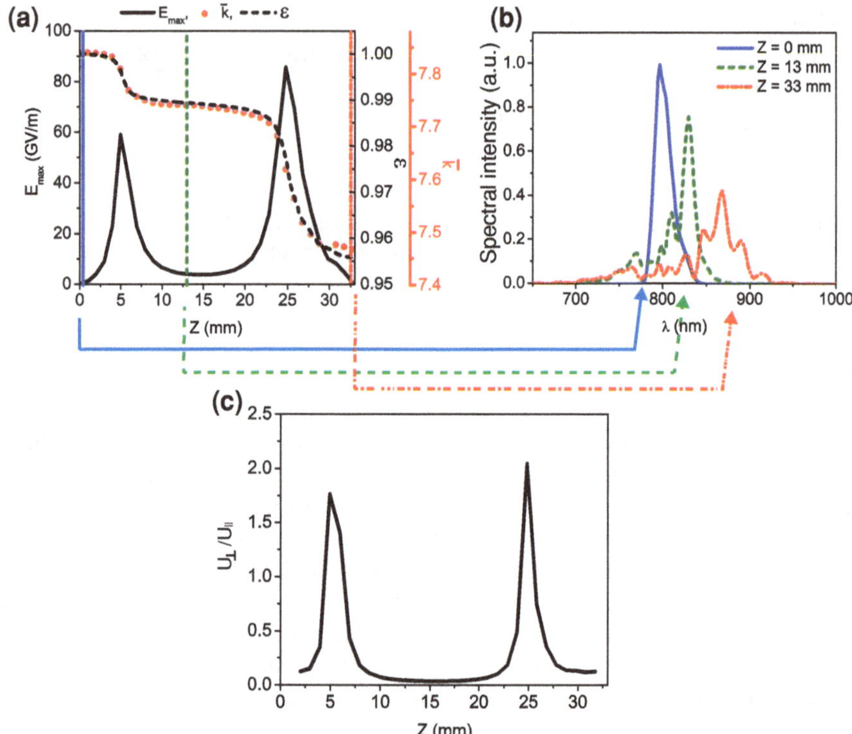

Fig. 6.13 Figures from [100]. PIC simulation result for a laser of $r_0 = 18\,\mu\text{m}$, $r_m = 41\,\mu\text{m}$, $a_0 = 1.2$, focused 4.8 mm into a 33 mm long capillary. **a** Peak accelerating field (*solid black*), laser energy (*dashed black*), and $\bar{k}$ (*red*), showing the amount of redshift as functions of propagation distance in the plasma channel. **b** Laser spectra before the channel (*solid blue*), after first E_{max} peak (*dashed green*) and at the exit of the channel (*dot-dashed red*). **c** The ratio of energy in the transverse and the longitudinal fields as defined by Eq. (6.9)

in the channel. The ratio was calculated using the fields for approximately three plasma periods behind the laser pulse. This partition of energy into the fields depends on the laser evolution. For the experimental parameters discussed in this article, laser spot size evolution leads to the wake excitation in both linear and nonlinear regimes, motivating the need for a numerical investigation.

Both one-dimensional theory (Sect. 6.4.2) and r–z PIC simulations illustrated the correlation between wakefield amplitudes and the redshift in a channel guided LPA. Section 6.4.5 will discuss detailed spectral analysis of experimental and simulation results.

6.4.5 Analysis of Optical Spectra

6.4.5.1 Dependence on a_0 and n_0

The dependence of ΔR on a_0 and n_0 were studied to understand the dynamics of wake excitation in the experiments. The temporal pulse shape and laser focus position were fixed. Experimental parameters are listed in Table 6.3 (data sets 1 and 2). Figure 6.14a shows example spectra for a_0 between 1.0 and 1.3 for $n_0 = 1.5 \times 10^{18}$ cm^{-3}. With increasing a_0, the spectra redshifted and broadened. Since the laser pulses arrived at the capillary after the hydrogen gas was fully ionized, the spectral blueshift was minimal for all the experimental data discussed in this article. For the data and simulations, ΔR was calculated using Eq. (6.6). Figure 6.14b shows increasing ΔR as a function of a_0^2, illustrating the increased efficiency of the laser energy transfer to the plasma at higher intensities. Simulated ΔR is also shown in Fig. 6.14b. The simulations were performed using measured a_0, measured temporal profile, and linear density ramps [profile (II) in Fig. 6.12b] as will be discussed in Sects. 6.4.5.2 and 6.4.5.3. Since there is ~20 % uncertainty on the calculation of n_0, simulated ΔR for two different n_0 are shown [60, 107]. For the measured n_0, simulations overestimate ΔR compared to the experimental data. However, decreasing n_0 in the simulation by only 20 % accounts for the difference. Example spectra from the density scans $n_0 = 1.1$–2.1×10^{18} cm^{-3} for $a_0 = 1.2$ are shown in Fig. 6.14c, illustrating the increased redshift with n_0. Figure 6.14d shows linear dependence of ΔR on $n_0^{3/2}$ for $a_0 = 1.0$ and 1.2 as expected by theory [Eq. (6.8)]. The simulation underestimates ΔR for $n_0^{3/2} < 1$ and overestimates for $n_0^{3/2} \sim 1.8$. One possible cause of this effect is the modeling of the longitudinal density ramps as will be discussed in Sect. 6.4.5. The ramp profiles influence ΔR significantly, and depending on how the profile changes with n_0, the ΔR dependence on $n_0^{3/2}$ can be different. Overall, the experimental data on ΔR indicated increased efficiency of laser energy transfer to plasma with increased a_0 and n_0, and experimental data agreed with simulations within uncertainties.

6.4.5.2 Dependence on Temporal Pulse Shape

In a Ti:Sapphire laser system based on chirped-pulse amplification (CPA), pulses with different temporal shapes can be generated by manipulating the laser chirp [104]. The basic physics of a CPA system was introduced in Chap. 2. While the instantaneous frequency chirp has minor effects on wake excitation, pulses with different temporal shapes can excite different magnitude wakes. A steeper intensity rise can excite a larger wake underneath the pulse envelope compared to a gentle rise. The larger density gradient under the pulse enhances the pulse compression, which in turn results in higher a_0 and larger wake amplitudes. The details of wake excitations and enhancement of electron injection by skewed pulses are discussed elsewhere [102, 103].

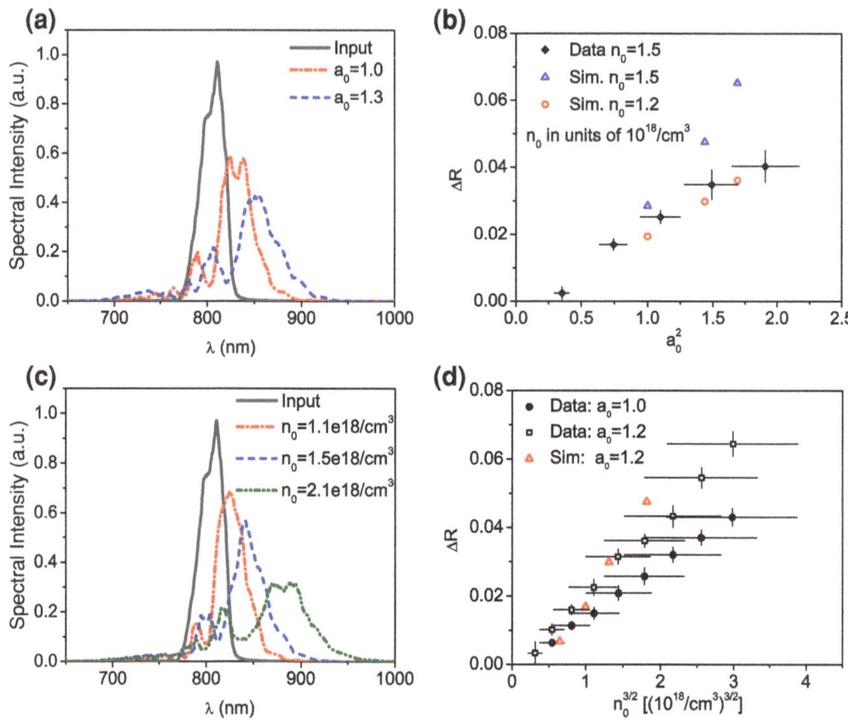

Fig. 6.14 Figures from [100]. **a** Normalized input and output optical spectra for input $a_0 = 1.0$ and 1.3. Each line is an average of ~15 shots. **b** ΔR as a function of a_0^2 for $n_0 = 1.5 \times 10^{18}$ cm^{-3}. ΔR from the experiment and simulations are shown. **c** Measured optical spectra for $n_0 = 1.1, 1.5, 2.1 \times 10^{18}$ cm^{-3} for the case of $a_0 = 1.2$. Each line is an average of ~15 shots. **d** ΔR as a function of $n_0^{3/2}$ for $a_0 = 1.0$ and 1.2. ΔR from the experiment and simulations are shown. Simulations were performed using measured a_0, measured temporal profile, and linear density ramps [profile (II)] as shown in Fig. 6.12b

The difference in temporal shapes caused a significant difference in ΔR even for the same values of a_0 and n_0. The spacing between compressor gratings was adjusted to compress or de-compress laser pulses in the experiment (data set 3 in Table 6.3). This adjustment corresponded not only to the change of pulse duration but also to pulse shape [104]. Measured pulse duration and ΔR as a function of compressor setting are shown in Fig. 6.15. The ΔR was asymmetric with respect to the pulse duration. This trend was investigated with simulations.

Input temporal intensity shapes for pulses of ~63 fs at FWHM from opposite sides of the maximum compression setting are shown in Fig. 6.16a and b. The pulses travel to the left on the graphs. The pulse shown in Fig. 6.16a has a pre-pulse at ~100 fs with a steep intensity rise for the main pulse and was measured at compressor position −0.19 mm. The intensity rise of the pulse shown in Fig. 6.16b is more gentle than the one shown in (a) and was measured at compressor position 0.11 mm. After the

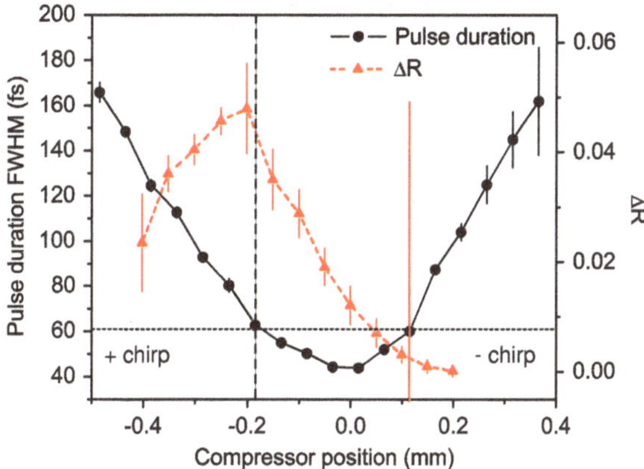

Fig. 6.15 Figures from [100]. Experimental results from a compressor scan. Pulse duration and ΔR as a function of compressor positions. The *vertical black dotted line* and the *red dot-dashed line* indicate where pulses shown in Fig. 6.16a and b were measured

laser-plasma interaction, the measured spectrum of the pulse with steep rise had a significantly greater ΔR than the gentle rise as shown in Fig. 6.16c. Simulation using these pulse shapes also resulted in a larger redshift for the pulse with the steep intensity rise [Fig. 6.16d], agreeing with the trend observed in the experiment. Simulated spectra for the opposite chirps are also shown in Fig. 6.16d. The simulations show that the frequency chirp did not change the trend, illustrating that the steep rise in the intensity profile is responsible for the increased ΔR. Even though the initial pulse durations measured at FWHM were the same, the ΔR differed greatly depending on the temporal shapes, which is consistent with previous results [102].

6.4.5.3 Dependence on Longitudinal Density Profile

Spectral analysis of experimental data (set 4 in Table 6.3) showed significant dependence of ΔR on laser focus location, Z_f, defined as the location of the vacuum laser focus with respect to the capillary entrance. For a given longitudinal density profile, the laser guiding condition changes with a different Z_f as discussed in Sect. 2.3.2. However, different longitudinal density profiles change how ΔR depends on Z_f. Two different longitudinal density ramps (shown in Fig. 6.12b) were studied to characterize their effects on ΔR as a function of Z_f and compared with experimental data.

In the experiments, gas inlets with 4 times larger cross sectional areas than the capillary were located 2 mm inside both ends as shown in Fig. 6.12b. Cold gas density simulation using ANSYS [113] predicted the electron density profile (I) in Fig. 6.12b. However, the plasma expansion following the capillary discharge can change the longitudinal electron density distribution. Since the gas in the inlets is not ionized, the plasma in the capillary can expand into the gas inlet regions and outside the capillary [107]. The combination of gas density distribution and plasma expansion can create a plasma density ramp that is longer than predicted by the cold gas density

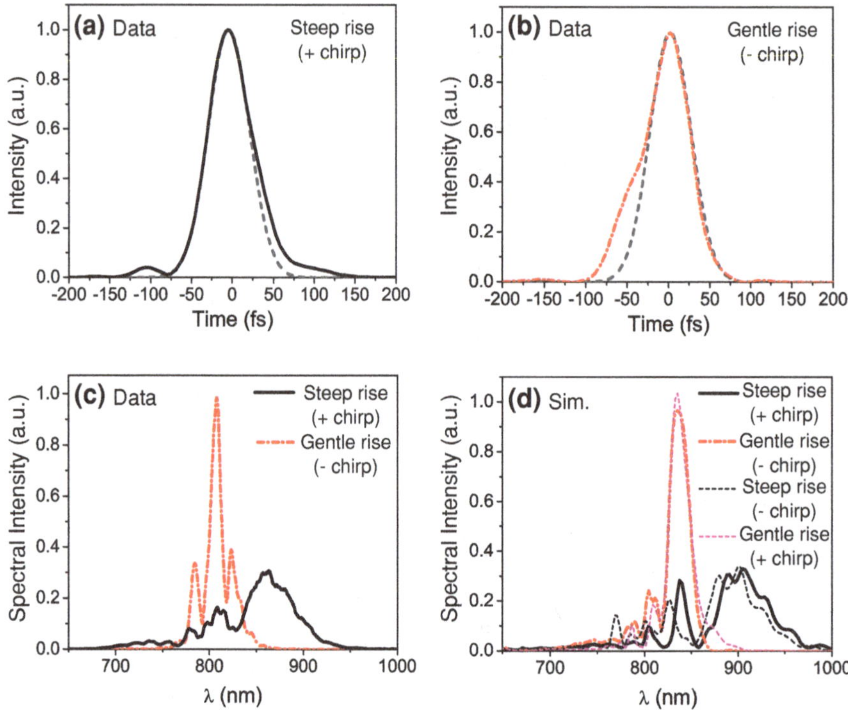

Fig. 6.16 Figures from [100]. Experimental and simulated results from a compressor scan. **a** and **b** Temporal pulse shapes measured by a GRENOUILLE at compressor positions −0.19 mm and 0.11 mm. The pulses travel to the left in the graphs and Gaussian distributions are shown for reference. **c** Experimentally measured spectra with a steep intensity rise at −0.20 mm and a gentle rise at 0.10 mm. **d** Simulated spectra using combinations of positive and negative chirps with the two intensity profiles (a) and (b)

simulation. We estimated this effect by using the ion acoustic shock expansion speed [114],

$$v_s = \sqrt{\frac{\gamma k_B T_e}{m_i}}, \tag{6.10}$$

where γ is the ratio of specific heats, k_B is the Boltzman constant, and m_i is the gas atomic mass. A temperature $T_e = 7\,\text{eV}$ and $\gamma = 5/3$ was used for the electron gas based on simulations performed by Bobrova et al. in Ref. [47]. Based on the transmission efficiency of the guided laser, the plasma channel was formed ~100 ns after the discharge. Assuming 100 ns of plasma expansion before the arrival of the laser, the shock would travel ~3.5 mm. Hence, a linear density ramp reaching 3.5 mm inside the gas inlet was constructed and shown as profile (II) in Fig. 6.12b.

The influence of the two ramps on ΔR were compared with experimental data. In the experiment, the laser focus position with respect to the entrance of the plasma

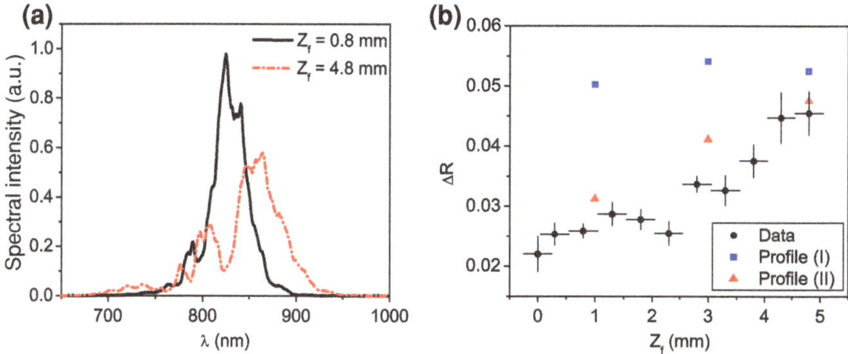

Fig. 6.17 Figures from [100]. **a** Measured optical spectra for $a_0 = 1.2$, $Z_f = 0.8$ mm (*solid black*) and $Z_f = 4.8$ mm (*dot-dashed red*). **b** ΔR as a function of Z_f. The larger Z_f indicates the laser is focused further downstream of the capillary entrance

channel was varied by moving the capillary along the laser axis. The averaged spectra and ΔR are shown in Fig. 6.17. A linear fit to the data indicates a ΔR increase of 0.004/mm, and the dependence is clearly visible in the spectra and ΔR. The simulated ΔR are also shown as blue squares and red triangles for profile (I) and (II) in Fig. 6.17b. With profile (I), the laser self-focused more strongly and reached higher E_{max} when focused further downstream than near the channel entrance. However, the laser guiding was better matched when focused near the channel entrance and excited wakes for a longer distance. As a result, this shift of Z_f did not change the total ΔR significantly. With profile (II), the simulated ΔR showed the same dependence on Z_f as the data, 0.004/mm, according to a linear fit. When the laser was focused at $Z_f = 1.0$ mm, the laser did not excite significant wakefields at the entrance of the channel due to the low plasma density. Additionally, focusing near the entrance of the channel did not improve guiding, resulting in much less ΔR than at $Z_f = 4.8$ mm.

This analysis shows that the density ramp at the ends of the capillary is a probable cause of the difference between experiments and simulations. Simulations using the long linear ramp [profile (II)] reproduced the trend observed in the experiment. Since the profile was an estimate, an experimental characterization of the density ramp would be valuable.

6.4.5.4 Estimate for Accelerating Field

The accelerating field was estimated using measured ΔR and simulations. Analysis thus far has illustrated that the efficiency of laser energy transfer to plasma was particularly sensitive to the temporal pulse shape and the longitudinal plasma density ramps. The effect of the higher order spatial profile on E_{max} was small for TREX pulses as discussed in Sect. 6.2. Therefore, a Gaussian transverse profile was assumed

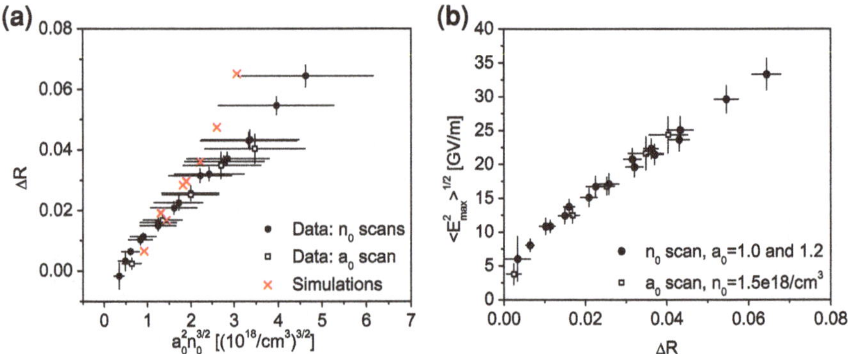

Fig. 6.18 Figures from [100]. **a** Measure of redshift, ΔR, as a function of vacuum $a_0^2 n_0^{3/2}$ for plasma density scans at $a_0 = 1.0$ and 1.2, and laser energy scan at $n_0 = 1.5 \times 10^{18}\,\mathrm{cm}^{-3}$ (*black*). Simulated results for $a_0 = 1.0$, 1.2, and 1.3 for $n_0 = 1.5 \times 10^{18}\,\mathrm{cm}^{-3}$, $n_0 = 1.2 \times 10^{18}\,\mathrm{cm}^{-3}$ and $n_0 = 0.75$–$1.5 \times 10^{18}\,\mathrm{cm}^{-3}$ for $a_0 = 1.2$ (*red*) are also shown. **b** Estimates of averaged wakefield amplitudes in the experiments based on ΔR. Calculated $\langle E_{\max}^2\rangle^{1/2}$ for the a_0 and n_0 scans (data set 1 and 2) are plotted. The density scans varied $n_0 = 0.7$–$2.1 \times 10^{18}\,\mathrm{cm}^{-3}$ for $a_0 = 1.0$ and 1.2. The energy scan varied $a_0 = 0.8$–1.3 for $n_0 = 1.5 \times 10^{18}\,\mathrm{cm}^{-3}$

and the experimentally measured temporal profile and long linear density profile [profile (II)] were implemented in the simulations.

Measured ΔR from a_0 and n_0 scans were plotted against $a_0^2 n_0^{3/2}$ in Fig. 6.18 to visualize the relationship described by Eq. (6.8). For these data sets, the temporal profile and Z_f were fixed, leading to a change in wake excitation due only to the change in a_0 and n_0. The graph shows that ΔR is consistent between scans of a_0 and n_0, illustrating that changing a_0 or n_0 changes the efficiency of the energy transfer as expected by $a_0^2 n_0^{3/2}$. The redshifts from simulated spectra are also shown in Fig. 6.18 and match experimental data within uncertainty for $a_0^2 n_0^{3/2} < 2.5$. Some possible causes of the difference at high $a_0^2 n_0^{3/2}$ are uncertainties on a_0, n_0, and longitudinal density profiles. However, assuming a given density profile and pulse shape, one can use ΔR to estimate the averaged accelerating field without knowing a_0 and n_0 precisely.

For each simulation in Fig. 6.18, ΔR and $n_0^{3/2}\langle E_{\max}^2/E_0^2\rangle$ were calculated independently. Equation (6.8) indicated $\Delta R \propto n_0^{3/2}\langle E_{\max}^2/E_0^2\rangle$ for matched guiding. The simulations showed deviation from this relation due to the oscillating spot size, resulting in wake excitation in both linear and nonlinear regimes. In the experimental regimes studied here, the simulations resulted in $\langle E_{\max}^2/E_0^2\rangle n_0^{3/2}[(10^{18}\,\mathrm{cm}^{-3})^{3/2}] = 31\,\Delta R^2 + \Delta R$. Using this relation, $\langle E_{\max}^2\rangle^{1/2}$ was estimated for the data and is shown in Fig. 6.18. It should be noted that experimentally measured a_0 and n_0 were different from those used in simulations for a given ΔR. The averaged $\langle E_{\max}^2\rangle^{1/2}$ ranged between 0 and 33 GV/m for $a_0 = 0.8$–1.3 and $n_0 = 0.7$–$2.1 \times 10^{18}\,\mathrm{cm}^{-3}$ in the experiments. This estimate is based on the total laser energy deposited in the plasma measured with the redshift, and the energy partition between the longitudinal and transverse fields calculated by the simulations.

6.4.6 Summary of Spectral Analysis

The analysis of spectral shifts to diagnose wakefields in a capillary discharged LPA was presented. This analysis was motivated by the need to better understand the accelerating field in an LPA to achieve controlled acceleration. The spectral redshifts are directly related to laser energy transfer in the plasma. The energy is used to excite the longitudinal and transverse wakefields, and the energy partition depends on the laser evolution in the channel. For the experimental conditions discussed in this article, the investigation of the relationship between the redshift and the accelerating field required numerical analysis. The energy transfer efficiency was found to be particularly sensitive to the temporal pulse shape and the focal location. The experimental conditions for these parameters were modeled and included in PIC simulations. These simulations were used to estimate the wake amplitudes in the experiments.

The spectral redshift showed increased efficiency of laser energy transfer into plasmas as functions of increasing a_0 and n_0. Moreover, the temporal pulse shape was shown to influence the efficiency of laser energy transfer into plasmas. The observed trend was reproduced when measured temporal pulse shapes were used in simulations, which showed that ΔR differed greatly due to the different pulse evolutions.

The effect of laser coupling conditions at the entrance of the channel was studied through laser focus position, Z_f, and longitudinal plasma density ramps at the ends of the channel. The experimental data showed significant dependence of ΔR on Z_f. Simulations showed that different longitudinal density profiles change the dependence of ΔR on Z_f. Applying a linear density ramp that accounts for the ion acoustic expansion of the plasma resulted in the simulations reproducing the experimental dependence of ΔR on Z_f.

The accelerating field was estimated using measured ΔR and simulations. Although the relationship between ΔR and E_{max} was complicated by the spot size evolution in the channel, ΔR is still related to E_{max} through Eq. (6.5) and measured ΔR scaled with $a_0^2 n_0^{3/2}$. The temporal profile and the linear density ramp were included in PIC simulations to achieve better agreement between the experiment and simulations. Using these simulations, averaged peak electric fields reaching up to 33 GV/m ($a_0 = 1.2$, $n_0 = 2.1 \times 10^{18}\,\text{cm}^{-3}$) were estimated for the experiments.

In conclusion, a detailed study of optical redshifting of drive laser light shows that it can be used as a diagnostic of laser energy transfer into plasmas. Quantitative agreement between experimental data and simulations can be obtained when proper input conditions of laser temporal profile and laser coupling condition (focal location and longitudinal plasma density ramps) are applied. The analysis implies that experimental characterization of the longitudinal density profile is important to quantify the wake excitation. Overall, the analysis demonstrated that optical spectra is a valuable diagnostic for laser energy depletion in plasma which helps us better understand the wake excitation below the e-beam production threshold.

6.5 Design Consideration for Staged LPA

6.5.1 Schemes for Multiple Laser Pulses

In the current staging experiment, timing jitter of the two driving laser pulses was not an issue since they originated from a single pulse. However, because the pulse was split after the amplification, each pulse was lower in energy, and laser pulse 2 underwent a severe temporal and spectral modulation during the splitting process. In this section, causes of the modulation, group velocity dispersion (GVD) and self-phase modulation (SPM), are introduced. These laser-matter interactions in the beam splitter reduced the peak intensity of laser pulse 2 and resulted in less efficient wake excitation in the 2nd module. For future experiments, these modulations should be minimized by a selection of the beam splitting optic and/or splitting prior to temporal compression to reduce intensity at the splitter. Furthermore, pulses could come from two different laser systems as will most likely be necessary for high energy accelerators. In this scenario, the timing jitter of the laser systems could be an issue to synchronize the pulses. This jitter could be minimized by using the same seed laser pulses and splitting the pulses before the amplification. The jitter will then be due to the difference in path lengths which can be compensated to tens of femtosecond level using a similar technology as those developed for pump-probe experiments in a light source [115, 116].

6.5.2 Group Velocity Dispersion

Pulse durations and shapes of laser pulse 2 were different from those of laser pulse 1. Laser pulse 2 undergoes GVD, which is a temporal stretch or compression caused by the wavelength dependent refractive index in the medium. The group velocity v_g in the medium is $v_g = c/N(\lambda)$ where the group index is $N(\lambda) = \eta(\lambda) - \lambda \mathrm{d}\eta(\lambda)/\mathrm{d}\lambda$ and $\eta(\lambda)$ is the refractive index. This is the velocity which the envelope of the pulse propagates in a medium assuming the absence of nonlinear effects. For an unchirped pulse of spectral width σ_λ, temporal broadening of a pulse is defined as $\sigma_\tau = |D_\lambda|\sigma_\lambda L$, where L is the length of the medium and $|D_\lambda| = -\lambda/c(\mathrm{d}^2\eta/\mathrm{d}\lambda^2)$. Most materials including fused silica are said to exhibit normal dispersion $D_\lambda < 0$, where longer wavelength components travel faster than the shorter wavelength components. The dispersion of fused silica is $D_\lambda = -107\,\mathrm{fs}/(\mu\mathrm{m}{-}\mathrm{mm})$ at $\lambda = 0.8\,\mu\mathrm{m}$. For an

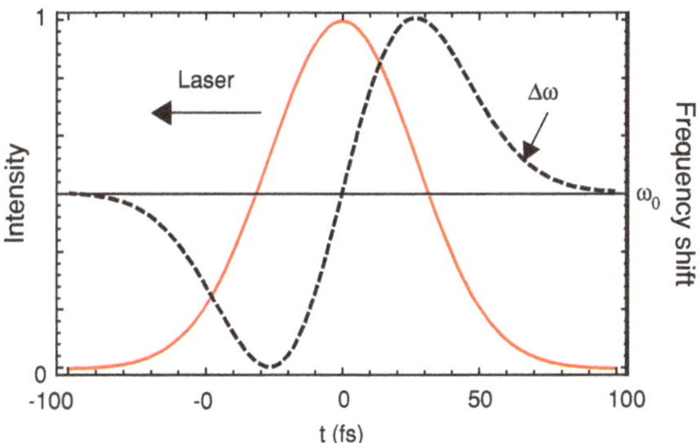

Fig. 6.19 Calculated frequency shifts due to the self-phase modulation. The leading edge of the pulse is frequency downshifted (*wavelength redshifted*) and the trailing edge is frequency upshifted (*wavelength blueshifted*)

initially transform limited (i.e. no chirp) pulse with $\sigma_T = 40\,\text{fs}$ propagating through the 10 mm fused silica can result in a positively chirped pulse with $\sigma_T = 47\,\text{fs}$.

6.5.3 Self-Phase Modulation

Optical spectra of laser pulse 2 were modulated from those of laser pulse 1 due to the SPM in the beam splitter. SPM is a spectral shift caused by a nonlinear polarization of the material induced by an intense laser field. When laser pulse 2 propagates through the beam splitter (laser pulse 1 reflects), the pulse experiences an intensity dependent refractive index, $\eta(I) = \eta_0 + \eta_2\, I$. The η_0 is the linear refractive index and η_2 is the second order nonlienar refractive index of the medium. The beam splitter is made of fused silica and $\eta_0 \approx 1.46$ and $\eta_2 \approx 2.5 \times 10^{-16}\,\text{cm}^2/\text{W}$ at $\lambda = 1.06\,\mu\text{m}$ [117]. As a result, the phase velocity of the light changes with intensity gradient, which leads to a wavelength redshifting ($\Delta\omega < 0$) at the leading edge and blueshifting ($\Delta\omega > 0$) at the trailing edge of the laser pulse as shown in Fig. 6.19. Effectively, the SPM introduces a positive chirp on laser pulse 2 propagating through the beam splitter. If the pulse is initially unchirped or positively chirped ($d\omega/dt > 0$), SPM can result in spectral broadening. If the pulse is initially negatively chirped ($d\omega/dt < 0$), SPM can lead to spectral compression.

6.5.4 Pulse Splitting in Staging Experiment

Laser pulse 2 undergoes an interplay of GVD and SPM simultaneously in the beam splitter. The SPM tends to redshift the leading edge and blueshift the trailing edge. The GVD leads to faster propagation of longer wavelength components than the shorter wavelength components. For an initially positively chirped pulse ($d\omega/dt > 0$), the output pulse is even more chirped and lengthened. For an initially negatively chirped

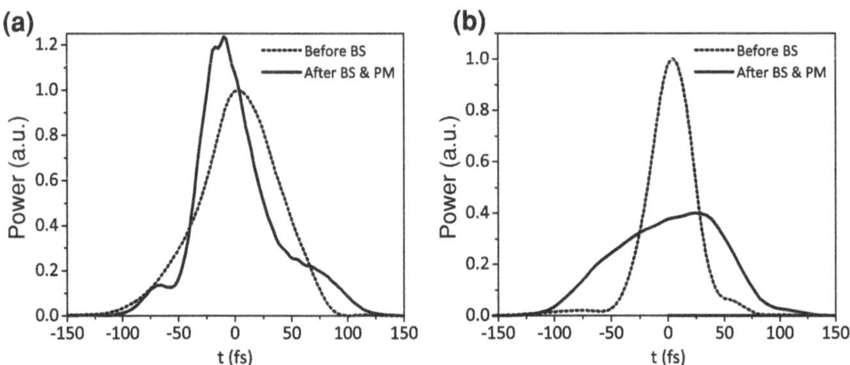

Fig. 6.20 Temporal intensity profiles measured before (*dashed curve*) and after (*solid curve*) the beam splitter and the plasma mirror for two compressor positions. **a** Compressor position of −0.077 which resulted in 76 fs before the beam splitter and 55 fs after the beam splitter and the plasma mirror at FWHM. **b** Compressor position of −0.277 which resulted in 42 fs before the beam splitter and ~130 fs after the beam splitter and the plasma mirror at FWHM

pulse ($d\omega/dt < 0$), spectral narrowing and temporal compression can be achieved. Figure 6.20 shows the measured pulse shapes before and after the beam splitter and plasma mirror for two compressor settings using GRENOUILLE [61]. The integrals of these curves are equated. The input pulse of Fig. 6.20a (dotted curve) was negatively chirped and was 76 fs at FWHM. The propagation in the beam splitter resulted in a pulse of 55 fs at FWHM (solid curve), demonstrating a temporal compression near the intensity peak. However, the pulse also has a significant broad pedestal such that the peak power was only 83 % of that of a Gaussian pulse with the same width. This condition was approximately the shortest pulse length for laser pulse 2. The input pulse of Fig. 6.20b was near the optimum compression and was 42 fs long at FWHM. The intense pulse was strongly modulated after the optics. The output pulse duration was ~130 fs. Although the GRENOUILLE retrieval errors were $\leq$0.011 (ideally $<$0.01) for all pulses shown here, the device is not optimized for the measurement of pulses longer than ~100 fs. The output pulse in (b) is shown to provide only the general trend. When laser pulse 2 was compressed as in Fig. 6.20a, a_0 for laser pulse 1 would be ~30 % lower than peak intensity achievable and would be more difficult to produce e-beams. When laser pulse 1 was compressed as in Fig. 6.20b, wake excitation in the 2nd module would be inefficient due to the lower a_0 of laser pulse 2. Since the compressor was shared between laser pulse 1 and laser pulse 2, the pulse lengths could not be optimized simultaneously for the two laser pulses. For the initial results on wake excitation in the 2nd module that will be presented in the following section, the pulse modulation interfered with efficient wake excitation.

The effect of SPM can be significantly reduced by making the beam splitter with a low nonlinear refractive index material such as magnesium fluoride (MgF_2). The nonlinear refractive index of MgF_2 is $\eta_2 \approx 7.6 \times 10^{-17}\,\text{cm}^2/\text{W}$ at $\lambda = 1.06\,\mu\text{m}$ [117]. The accumulated nonlinear phase shift in the beam splitter can be reduced to roughly 30 % of that induced in fused silica with a given intensity and thickness of the optic. The SPM can be further reduced if the laser pulses are split before the compressor gratings, reducing the laser intensity at the beam splitter. In this case, the two laser pulses could be compressed to ~45 fs simultaneously, allowing each stage to be operated at near the peak efficiency.

6.6 Experiment on Wake Excitation in 2nd Module

In this section, experiments on wake excitation in the 2nd module are presented. The goal of the experiment was to drive the 2nd module with laser pulses reflected off the plasma mirror and estimate the electron energy gain for an e-beam injected from the 1st module. The basic physics of wake excitation and details of the plasma mirror were presented in Chap. 2 and in Chap. 5, respectively. Because laser pulse 2 was temporally modulated through propagation in the beam splitter, pulse duration was longer than in the experiments discussed earlier in this chapter. While only spectral redshift was used as a measure of wake excitation in the experiments presented in Sect. 6.4.2, for a long pulse, both spectral blue and redshifts are measures of wake excitation. The schematics for this process are shown in Fig. 6.21. Laser pulses exciting wakefields dominantly undergo redshift when the pulse length is short compared to the plasma wavelength $c\Delta\tau \ll \lambda_p$ as shown in Fig. 6.21a. This is because the majority of the laser pulse is on a negative electron density gradient where the difference in phase velocity caused by the plasma density difference lengthens the wavelength. When the pulse is long as in this experiment ($\lambda_p \approx 19\,\mu\text{m}$ and $c\Delta\tau \approx 39\,\mu\text{m}$), the spectra can be both blueshifted and redshifted as shown in Fig. 6.21b. The trailing edge of the pulse experiences an increasing plasma density gradient which compresses the wavelength. As a result, both blue and redshifting are observed in post-interaction laser spectra.

6.6.1 Experimental Configuration

The experimental setup was described in detail in Chap. 3. Laser pulse 1 was blocked and laser pulse 2 was used as the drive laser. The energy of laser pulse 2 was ~650 mJ. This pulse reflected off the plasma mirror at 300 J/cm^2, achieving a reflectivity of ~80 % and resulting in ~520 mJ on the 2nd module. The Strehl ratio of the beam reflected off the plasma mirror was 0.7. Experiments were performed using a capillary of same length and diameter, 33 mm and 250 µm, as the one used in the spectra-based wakefield diagnostic experiment presented in Sect. 6.4. The laser parameters were $r_0 = 22\,\mu\text{m}$, $\Delta\tau \sim 130\,\text{fs}$, $a_0 \simeq 0.4$ and pulses arrived 50 ns before thepeak of

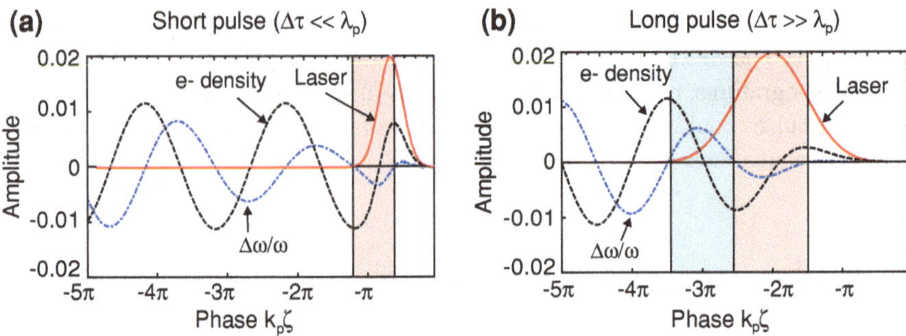

Fig. 6.21 Schematics of spectral shifts from plasma wave density gradient. **a** For a short pulse compared to the plasma wavelength, the laser is dominantly redshifted. **b** For a long pulse, the laser can be both redshifted and blueshifted

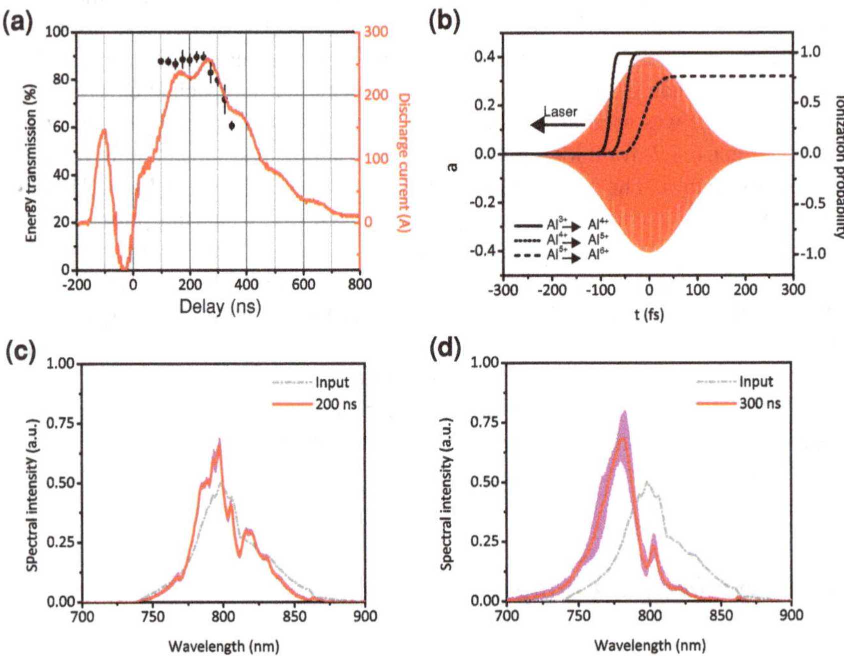

Fig. 6.22 **a** Laser energy transmission as a function of discharge delay with drive laser of $a_0 \sim 0.4$. The plasma channel was $n_0 \sim 2 \times 10^{18}\ \mathrm{cm}^{-3}$. **b** Theoretical aluminium ionization probabilities by a laser pulse $a_0 = 0.4$ and duration 130 fs at FWHM, same as laser pulse 2. **c** and **d** Measured input and output spectra from the scan in (**a**) at a 200 ns and 300 ns after the discharge initiation

discharge current which peaked at 250 A and 450 ns long at FWHM. Transmitted laser energy, laser mode and optical spectra were measured to optimize the guiding. The spectra were analyzed to compare with simulations and diagnose the excited wakefield.

6.6.2 Results and Analysis

Laser energy transmission and output modes were observed as a function of discharge timing to optimize the guiding similar to the experiments presented in Sect. 6.3. For the 1st module, the laser guiding was optimized with low power laser pulses. In this 2nd module experiment, the guiding was optimized at full power because the plasma mirror performance depended on the fluence on target as was discussed in Chap. 5. Optimization with low power laser pulses would not be the same as that performed with high power pulses. For this data, the plasma channel was $n_0 \sim 2\times10^{18}\,\text{cm}^{-3}$ and $r_\text{m} \sim 37\,\mu\text{m}$. Laser energy transmission as a function of discharge timing is shown in Fig. 6.22a. The transmission was calculated from measurements of total energy transmission and the assumption that plasma mirror reflectivity was 80 %. High guiding transmissions of ~90 % were observed for 100–250 ns after the initiation of the discharge. After 250 ns, the transmission rapidly decreased, probably due to an increased ionization rate of aluminium (Al) and oxygen (O) ablated from the sapphire (Al_2O_3) wall. For the intensities used in this experiment, the laser could have ionized to Al^{6+} and O^{6+} near the peak of the pulse. The ionization probabilities for Al are shown in Fig. 6.22b with the laser pulse $a_0 \sim 0.4$. Another observation suggesting the presence of ionization is the spectra. Input (dot-dashed grey curve) and output (solid red curve) laser spectra for 200 and 300 ns are shown in Fig. 6.22c and d. The absence of significant redshift indicates that there is no significant wake excitation in the plasma channel. The later delay of 300 ns shows a significant spectral blueshift compared to the input spectrum. This is likely due to ionization induced blueshift where the increasing plasma density gradient due to the ionization causes spectral blueshift [51]. A previous experiment by Spence et al. also showed ablation of capillary walls by discharge current of 300 A peak and 200 ns FWHM [106]. They estimated the plasma electron increase caused by the ablation to be <2.5 %. In our experiment, the discharge current was much longer (450 ns at FWHM) which may have increased the ablation. To mitigate this effect, recent improvements to the discharge circuitry have lowered the drive current amplitude and pulse duration. The ionization by driving laser pulses can increase plasma density as well as defocus the laser pulse. When the laser intensity is such that the ionization rate varies as a function of transverse position r (due to the transverse laser profile), electron density can peak on axis (and minimum refractive index on axis) so as to defocus the laser pulse [118]. Since this effect in wake excitation is yet to be fully characterized, for the following wake excitation experiments, the delay of 200 ns was chosen to minimize this effect.

Good laser guiding through the 2nd module using pulses reflected off the plasma mirror was demonstrated. Examples of input and output modes are shown in Fig. 6.23. The input mode (a) was the reflection off the plasma mirror with a Strehl ratio 0.7. The laser mode at the exit of the 2nd module is shown in (b). The density in the plasma channel was increased to $n_0 \sim 3.1 \times 10^{18}\,\text{cm}^{-3}$ and $r_\text{m} \sim 34\,\mu\text{m}$ to compensate for the lower a_0 compared to the experiment discussed in Sect. 6.4. The guided laser mode was symmetric and the total laser energy transmission was 73 %. Assuming

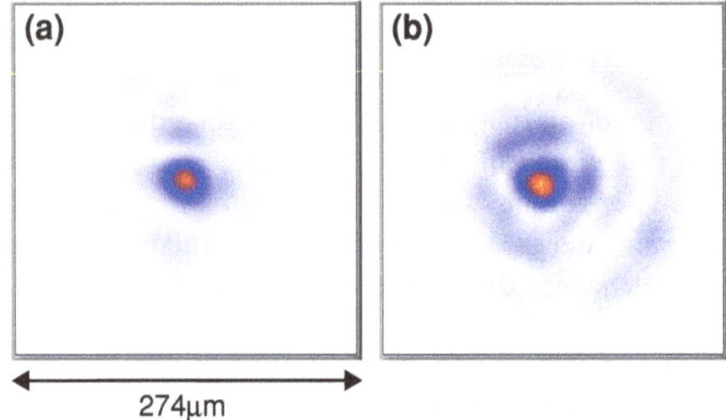

Fig. 6.23 Examples of input and output laser modes in the 2nd module. **a** Input laser mode which was reflected laser pulse at 300 J/cm^2 from the plasma mirror. **b** Laser mode at the exit of the capillary using $n_0 \sim 3.1 \times 10^{18}\,\text{cm}^{-3}$ and $r_\text{m} \sim 34\,\mu\text{m}$ plasma channel

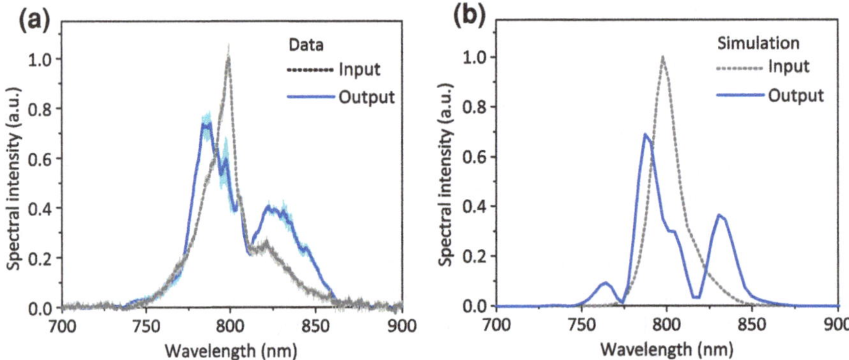

Fig. 6.24 **a** Input (*dotted grey*) and output (*solid blue*) laser spectra guided through the 2nd module. The plots are average of ~5 shots. Standard deviations are shown as *grey and light blue lines*. **b** Simulated input (*dotted grey*) and output (*solid blue*) spectra for the experimental parameters

80 % plasma mirror reflectivity, the guiding transmission was 91 %, indicating a good laser guiding.

Compared to the experiments discussed in Sect. 6.4, the amount of redshift was minimal (see Fig. 6.14 for comparison) suggesting that the amplitudes of the wakefields were small. The averaged spectra of input pulse (grey curve) and guided pulse (blue curve) through a plasma channel of $n_0 \sim 3.4 \times 10^{18}\,\text{cm}^{-3}$ and $r_\text{m} \sim 34\,\mu\text{m}$ are shown in Fig. 6.24a. The standard deviations are indicated with light grey and light blue lines. Because of the significant blueshift, the central wavelength shift in the output spectra was only $\Delta R \sim 0.003$ where ΔR was defined in Eq. (6.7).

Simulations were performed to compare with data and reproduce qualitative spectral features. They showed a small wakefield amplitude. Measured input pulse shape and spectra were used in INF&RNO PIC code along with n_0, r_m and a_0 used in the experiment. The longitudinal plasma density was modeled with a linear density ramp reaching 3.5 mm inside the gas inlet shown as Profile (II) in Fig. 6.12b. This was reasonable since laser pulses arrived ~200 ns after the discharge begun as in the experiment discussed in Sect. 6.4. The input spectra (grey curve) and the output spectra (blue) are shown in Fig. 6.24b. Data and simulation are qualitatively similar, showing both blueshifts and redshifts. The spectral redshift was similar to the data and $\Delta R \sim 0.002$. According to the simulation, the energy gain of an electron traveling at the speed of light in the plasma wave is ~13 MeV. The analysis discussed in Sect. 6.4.2 showed that different temporal pulse shapes can result in different wakefield amplitudes, and hence the spectral redshifts. Since laser pulse 2 was strongly modulated, the simulations provided a qualitative sense of field amplitudes.

Simulations indicated that the electron energy gain could be ~100 MeV if the laser pulses were compressed to $\Delta\tau = 55$ fs. Two simulations were performed to gain insights into wake excitation at the laser pulse 2 energy level. The first one was similar to the data discussed above ($r_0 = 22\,\mu$m, $\Delta\tau = 130$ fs, $a_0 = 0.4$, $n_0 = 3 \times 10^{18}\,\text{cm}^{-3}$ and $r_m = 34\,\mu$m), but using a Gaussian temporal profile. The electron energy gain as a function of phase ($\zeta = z - ct$) and propagation distance is shown in Fig. 6.25a. The laser propagates near $\zeta \sim 0$ and the hashed region indicates the defocused region. The vertical axis is the laser propagation distance in the channel where $Y = 0$ is the capillary entrance. With this long pulse, the energy gain would only be ~10 MeV, close to the previous simulation. If the same laser pulse were to be compressed to $\Delta\tau = 55$ fs ($a_0 \sim 0.6$) and a shorter channel of 15–20 mm were employed, the energy gain could be ~100 MeV. Since the dephasing length is $L_d \sim 6$ mm for $a_0^2 \ll 1$, the e-beam loses energy when the channel is too long (i.e. 33 mm). With the current experimental setup, 100 MeV energy gain may be difficult because the pulse shape at the compressed position is not a Gaussian profile but has a broad pedestal as seen in Fig. 6.20a. The energy gain would be ~15 % lower than that of a Gaussian pulse with the same duration assuming the E_{max} described in Eq. (2.22). If the SPM in the beam splitter were mitigated, a larger energy gain could be possible.

6.7 Summary and Conclusions

Wakefield excitation by a Gaussian beam and the LG beam were compared using simulation. TREX laser pulses were characterized by higher order LG beams based on a wavefront measurement which was then used as a drive laser in the simulation. While the longitudinal field E_{max} was similar for the two beams, differences in transverse field amplitude and size were observed. For experiments using TREX laser pulses, the study suggests that consideration of higher order modal content would be important in investigation of the e-beam properties such as charge and divergence.

An improved experimental technique to characterize the channel depth r_m and shape using laser centroid oscillation in the channel was demonstrated. The study was motivated by the need to better understand the dynamically evolving plasma

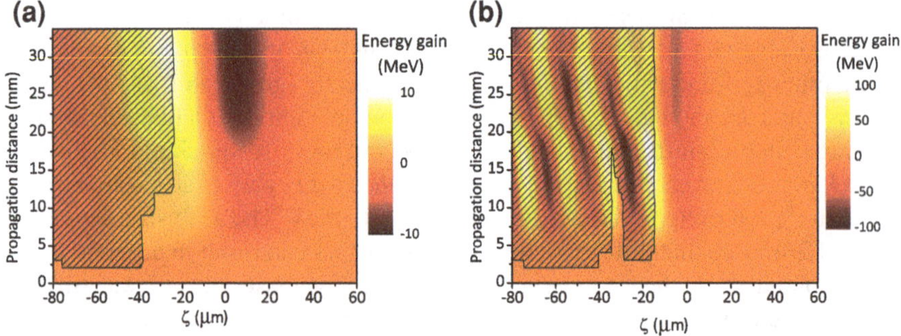

Fig. 6.25 Simulated electron energy gain as a function of phase ($\zeta = z - ct$) and propagation distance. The laser propagates near $\zeta \sim 0$ and the hashed region indicates the defocused region. The vertical axis is the laser propagation distance in the channel where $Y = 0$ is the capillary entrance. Laser was focused 3.5 mm into the plasma channel which was $n_0 = 3 \times 10^{18}\,\text{cm}^{-3}$ and $r_m = 34\,\mu\text{m}$. **a** The simulation for a long pulse similar to the case discussed above ($a_0 = 0.4$ and $\Delta\tau = 133\,\text{fs}$), indicating energy gain of ~10 MeV. **b** The simulation for a more compressed pulse ($a_0 = 0.6$ and $\Delta\tau = 55\,\text{fs}$), indicating energy gain ~100 MeV at ~15 mm propagation distance. Due to the dephasing, e-beam loses energy when the propagation distance is longer than ~15 mm

channel. The measured r_m agreed well with simulations, and the channel profile was inferred to be parabolic for $r < r_m$. The centroid oscillations also allowed a more precise alignment of the channel with respect to the laser, which will be critical for staged LPAs to transport e-beams between stages and post-accelerate efficiently.

The wakefield was diagnosed using optical spectral redshift and simulations. Single shot, non-destructive methods to gain insights into the excited wakefield and consequently the electron energy gain will be crucial for staged LPAs. Quantitative agreement between experimental data and simulations was obtained when proper input conditions of laser temporal profile and laser coupling condition were applied. The average accelerating gradient was then estimated from simulation based on measured spectral redshift.

The intense laser pulse splitting process which introduced the modulations and reduced the peak intensity of laser pulse 2 in the staging experiment was discussed. Self-phase modulation and group velocity dispersion were discussed as causes of this modulation. Mitigation of self-phase modulation using a beam splitter made of MgF_2 was discussed. The nonlinear phase shift would be reduced to 30 % of that from the current optic.

Initial experimental results on wake excitation in the 2nd module were presented. Good laser guiding using pulses reflected off the plasma mirror was demonstrated with 90 % energy transmission and symmetric output laser modes. Measured optical spectra were compared with simulation, and excitation of small amplitude wakefields were inferred. The energy gain of an e-beam would be ~10 MeV. If the SPM were mitigated and the laser pulse was compressed to 55 fs, the energy gain could be as high as 100 MeV, allowing easier demonstration of post-acceleration in the 2nd module.

It was presented in Chap. 4 that the initial e-beam production experiment demonstrated energy fluctuation of ~31 MeV at peak energy ~240 MeV with an energy spread ~35 MeV using ionization injection in the jet + cap target. With this e-beam, the current expected energy gain of ~10 MeV in the 2nd module would not be sufficient to demonstrate staged acceleration. If the SPM were reduced and an energy gain of ~100 MeV can be expected from the 2nd module, the demonstration of staged acceleration is likely possible.

Chapter 7
Summary and Conclusions

This thesis investigated experimental aspects of staging two LPAs. Tajima and Dawson proposed creating plasma density oscillations (plasma waves) through the interaction of an intense laser pulse with plasma [13]. The electric field sustained by the plasma wave can be tens of GV/m, providing a high acceleration gradient for electrons. However, there is an optimal acceleration length and consequently electron energy gain from a single stage LPA. Physics limiting the optimal acceleration length include laser diffraction, electron dephasing and pump depletion [27]. If the laser is guided to mitigate the diffraction, and the electron dephasing is controlled, energy gain is limited by depletion. For an acceleration to higher energy than can be achieved within a depletion length, or an acceleration of e-beam while maintaining control over electron bunch properties, LPA designs will rely on staging. Hence, the demonstration of staged acceleration is a critical step in the development of LPAs. This thesis focused on the following aspects:

I. Experimental investigation of e-beam production and characterization from the 1st module. In a staged LPA, production of a high quality e-beam is more critical than ever since the e-beam must be transferred to the sequential stage and accelerated in a controlled manner.
II. Development and characterization of a plasma mirror. The plasma mirror is a critical component in the staging experiment which allows for a compact coupling of intense laser pulses. This was the first plasma mirror using VHS tape.
III. Experiments on wake excitation and characterization below the e-beam production threshold for the 2nd module. Controlled wake excitation and a diagnostic of accelerating field is critical to estimate the energy gain of externally injected e-beams.
IV. Discussion on e-beam transport including emittance preservation between the stages, e-beam matching to a plasma wave in the 2nd module, and e-beam interaction with a plasma mirror.

The on-going experiment at the LOASIS Program to demonstrate staging was introduced in Chap. 3. During the period of my study, the staged acceleration beamline was designed, built and commissioned. The beamline that splits TREX laser

S. Shiraishi, *Investigation of Staged Laser-Plasma Acceleration*,
Springer Theses, DOI: 10.1007/978-3-319-08569-2_7

pulses into two to drive the two stages and diagnose the post-interaction laser pulses and e-beams was presented. Experimental results on e-beam production, plasma mirror, and wake excitation below the electron trapping threshold were presented in Chaps. 4–6 and summarized below.

Electron beam production was experimentally investigated in the undulator beamline. Pulses from the TREX laser system were used as drive laser pulses with the same focusing geometry and similar LPA targets as the staging experiment. These experiments allowed investigation of e-beam production with a greater laser energy range in a well established experimental setup. Injection using self-trapping and ionization in a plasma channel produced e-beams with large energy spread and large energy fluctuations. Electron beam quality was improved when the plasma density was tailored using a capillary waveguide with a embedded gas jet. The high density region from the gas jet self-focused the laser to achieve higher a_0. The increase in a_0, along with the negative density gradient which reduced the phase velocity of the wake, facilitated e-beam injection. Since injection was localized near the gas jet, e-beams with narrow energy spread $\sigma_{\Delta E/E} \simeq 5\,\%$, peak energy around 340 MeV with $\sigma_E \simeq 1.9\,\%$ fluctuation were injected at $\sim$100 % probability. This result was published in Gonsalves et al. [23]. This e-beam quality would be suitable for the staging experiment.

Slice energy spread and emittance were measured for e-beams produced from LPAs using the same focusing geometry and similar capillaries as the staging experiment. Simulations indicate that the measured energy spread of e-beams are dominated by correlated spread, which can be removed by placing the e-beam in the correct phase of the accelerating field. However, the slice energy spread is an intrinsic energy spread. Laser induced momentum modulations in e-beams were observed using coherent optical transition radiation (COTR). The coherent enhancement of OTR suggested an energy slice spread of $\Delta\gamma/\gamma \leq 0.5\,\%$ and the result was published in Lin et al. [64]. Furthermore, the emittance of e-beams produced from an LPA was estimated using an e-beam size measurement based on X-ray spectra and divergence measured using a magnetic spectrometer. An example e-beam of measured transverse size $\sigma \simeq 0.1\,\mu$m and divergence 1 mrad was presented. In the experiments, the measured emittance ranged from 0.1 to 1 mm-mrad for self-trapped e-beams and the result was published in Plateau et al. [68]. The e-beam characterization is critical for the design of future staged LPAs.

Based on the studies presented above, e-beam production in the 1st module was investigated in the staging beamline. A combination of the ionization injection and the tailored density profile was used to lower the injection threshold and stabilize e-beam properties. Electron beams with energy spread $\sigma_{\Delta E/E} \simeq 16\,\%$, peak energy $\sim$240 MeV and $\sigma_E \simeq 14\,\%$ fluctuation were observed with injection probability of $\sim$73 %. This e-beam would be sufficient when $\sim$100 MeV energy gain in the 2nd module can be expected. Since lower energy gain is predicted for the initial staging experiment, a re-design of the density profile in the 1st module is in progress to further improve the stability.

The tape-drive based plasma mirror was developed and characterized. The maximum reflectivity of $\sim$80 % and Strehl ratio of 0.7 were observed at a fluence of 300–700 J/cm^2 on the tape. To achieve a pointing stability comparable to the inherent laser pointing stability of the system, a tape monitor camera was implemented to assist in confirming surface quality. As a result, a plasma mirror with sufficient reflective quality for the staging experiment was demonstrated.

The effect of higher order profiles on wakefields in TREX experiments were investigated. Wakefield excitation by a Gaussian beam and the LG beam modeled based on measured wavefront were compared using simulation. While the longitudinal field E_{max} was similar for the two beams, differences in transverse field amplitude and shapes were observed. Drive laser pulses also evolved differently for the two profiles. This analysis was a part of publication, Shiraishi et al. [100]. The study suggested that consideration of higher order modal content is important for investigations of e-beam properties such as charge and divergence for future LPAs.

An improved experimental technique to characterize a plasma channel depth r_{m} and shape using laser centroid oscillation in the channel was presented. The measured r_{m} agreed well with simulations, and the channel profile was inferred to be parabolic for $r < r_{\mathrm{m}}$. The centroid oscillations also allowed a more precise alignment of the channel with respect to the laser, which will be critical for staged LPAs to transport e-beams between stages and post-accelerate efficiently. The result was published in Gonsalves et al. [105].

Amplitudes of wakefields were diagnosed using optical spectral redshift and simulations. Non-destructive methods to gain insights into the excited wakefield and consequently the electron energy gain will be crucial for staged LPAs. Quantitative agreement between experimental data and simulations was obtained when proper input conditions of laser temporal profile and laser coupling condition were applied. The average accelerating gradient was then estimated based on simulation and measured spectral redshift. For the data set used in this analysis, averaged peak electric fields reaching up to 33 GV/m ($a_0 = 1.2$, $n_0 = 2.1 \times 10^{18}/\mathrm{cm}^3$) were estimated, and the result was published in Shiraishi et al. [100]. This experiment demonstrated an excitation of a large accelerating field with the same capillary geometry as the 2nd module in the staging experiment. Similar spectral analysis was then used to diagnose wakefield amplitude in the 2nd module.

Initial experimental results on wake excitation in the 2nd module were presented. Good laser guiding using pulses reflected off the plasma mirror was demonstrated with 90 % energy transmission and symmetric output laser modes. Measured optical spectra were compared with simulation, and the e-beam energy gain of ∼13 MeV was inferred from simulations. The self-phase modulation and group velocity dispersion in the beam splitter which reduced the peak intensity of laser pulse 2 was discussed as a cause of the small wakefield amplitude in the 2nd module. If these effects were mitigated using a beam splitter made of MgF_2, the electron energy gain in the 2nd module could be as high as 100 MeV. This will allow an easier demonstration of post-acceleration in the 2nd module.

Preservation of e-beam quality between the stages were explored theoretically. The emittance of the e-beam can degrade if the e-beam spot size is small, divergence is large, and energy spread is large. As the e-beam increases in energy, the emittance growth will reduce. This effect is particularly significant in this staging experiment since e-beam energy is only 200–300 MeV. The mitigation of the emittance growth can be achieved through decreasing the focusing force via plasma density downramp, possibly expanding the laser spot size as is the case for mismatched laser guiding or by choosing a stage energy of the 1st module to achieve higher energy. To capture the e-beam and post-accelerate efficiently in the 2nd module, the transverse size of the e-beam σ_x and the matched spot size of the plasma wave σ_{m} need to be matched. This may be achieved when the e-beam is focused between the stages.

Furthermore, the effect of a plasma mirror on the e-beam was evaluated. The dynamical coupling of plasma and e-beam is expected to be small because the charge density in the e-beam is expected to be significantly less than that in the plasma mirror. Electron-ion scattering of e-beams propagating in an un-ionized polyester tape and a fully ionized sheet of plasma were evaluated. For the e-beam energy level in the staging experiment, the scattering is expected to increase the divergence of e-beams by a few hundred microradians. The scattering will be significantly reduced for a higher energy or a narrower divergence e-beam as in a 10 GeV stage LPA [28]. These discussions provided a rough estimate of e-beam and plasma mirror interactions in staged LPAs.

In conclusion, this thesis investigated critical physics for staged LPAs. The staging experimental beamline and the target system have been developed and commissioned at the LOASIS facility. Reliable e-beam production and efficient wake excitation necessary for the staged acceleration have been independently demonstrated with 40 TW pulses in the undulator beamline. Adjustment of the 1st module geometry to achieve stable e-beam production using the lower laser power, 25 TW, in the staging setup is in progress. The coupling of laser pulse 2 into the 2nd module using the tape-drive based plasma mirror was successfully achieved. Improvement of the wake excitation in the 2nd module via mitigation of laser pulse 2 modulation in the beam splitter is in progress. When reliable e-beam production and wake excitation in the staging setup are achieved, the staged acceleration experiment is ready to begin. The anticipated proof of principle demonstration of staged acceleration will be an exciting beginning for LPAs to further develop towards the next generation of compact high energy accelerators.

Curriculum Vitae

Satomi Shiraishi received a B.S. in Mathematics and a B.A. in Physics with Honors from the University of Chicago in 2007. Her undergraduate research was in experimental particle physics where she set a direct bound on the total decay width of the Top quark using data collected at the Collider Detector at Fermilab. As a graduate student in the group of Professor Young-Kee Kim at the University of Chicago, she spent 4 years at the LOASIS facility at Lawrence Berkeley National Laboratory under the supervision of Dr. Wim Leemans. Her work at LOASIS focused on diagnosing laser-plasma interactions and conducting experiments that form the basis of staged laser-plasma accelerators. She received the Gaurang and Kanwal Yodh Prize for outstanding experimental work from the University of Chicago and graduated with her Ph.D. in December 2013.

S. Shiraishi, *Investigation of Staged Laser-Plasma Acceleration*,
Springer Theses, DOI: 10.1007/978-3-319-08569-2

References

1. Lee, S.Y.: Accelerator Physics, 2nd edn. World Scientific, Danvers (2004)
2. Edwards, D.A., Syphers, M.J.: An Introduction to the Phyiscs of High Energy Accelerators. Wiley, New York (1993)
3. Particle Accelerator Physics, vol. 1, 2nd edn. Springer, Berlin (2004)
4. International Linear Collider Reference Design Report. ILC Global Design Effort and World Wide Study (2007)
5. Esarey, E., Schroeder, C.B., Leemans, W.P.: Physics of laser-driven plasma-based electron accelerators. Rev. Mod. Phys. **81**(3), 1229–1285 (2009)
6. Mangles, S.P.D., et al.: Monoenergetic beams of relativistic electrons from intense laser-plasma interactions. Nature **431**, 535–538 (2004)
7. Faure, J., et al.: A laser-plasma accelerator producing monoenergetic electron beams. Nature **431**, 541–544 (2004)
8. Geddes, C.G.R., et al.: High-quality electron beams from a laser wakefield accelerator using plasma-channel guiding. Nature **431**, 538–541 (2004)
9. van Tilborg, J., et al.: Temporal characterization of femtosecond laser-plasma-accelerated electron bunches using terahertz radiation. Phys. Rev. Lett. **96**(13), 014801 (2006)
10. Esarey, E., et al.: Overview of plasma-based accelerator concepts. IEEE Trans. Plasma Sci. **24**(2), 252–288 (1996)
11. Hooker, S.M.: Developments in laser-driven plasma accelerators. Nat. Photonics **7**, 775–782 (2013)
12. Gibbon, P.: Short Pulse Laser Interactions with Matter. Imperical College Press, London (2007)
13. Tajima, T., Dawson, J.M.: Laser electron accelerator. Phys. Rev. Lett. **43**(4), 267 (1979)
14. Strickland, D., Mourou, G.: Compression of amplified chirped optical pulses. Opt. Commun. **56**(3), 219 (1985)
15. Leemans, W.P., et al.: GeV electron beams from a centimetre-scale accelerator. Nat. Phys. **2**, 696 (2006)
16. Nakamura, K., et al.: GeV electron beams from a centimeter-scale channel guided laser wakefield accelerator. Phys. Plasmas **14**(05), 056708 (2007)
17. Wang, X., et al.: Quasi-monoenergetic laser-plasma acceleration of electrons to 2 GeV. Nat. Commun. **4**, article no. 1988 (2013)
18. Leemans, W.P., et al. In preparation
19. Bulanov, S., et al.: Particle injection into the wave acceleration phase due to nonlinear wake wave breaking. Phys. Rev. E **58**(5), 5257 (1998)
20. Faure, J., et al.: Controlled injection and acceleration of electrons in plasma wakefields by colliding laser pulses. Nature **444**, 737–739 (2006)

S. Shiraishi, *Investigation of Staged Laser-Plasma Acceleration*,
Springer Theses, DOI: 10.1007/978-3-319-08569-2

21. Liu, J.S., et al.: All-optical cascaded laser wakefield accelerator using ionization-induced injection. Phys. Rev. Lett. **107**(3), 035001 (2011)
22. Kaganovich, D., et al.: Measurements and simulations of shock wave generated plasma-vacuum interface. Phys. Plasmas **18**(12), 120701 (2011)
23. Gonsalves, A.J., et al.: Tunable laser plasma accelerator based on longitudinal density tailoring. Nat. Phys. **7**, 862–866 (2011)
24. McGuffey, C., et al.: Ionization induced trapping in a laser wakefield accelerator. Phys. Rev. Lett. **104**, 025004 (2010)
25. Chen, M., et al.: Theory of ionization-induced trapping in laser-plasma accelerators. Phys. Plasmas **19**, 033101–033104 (2012)
26. Leemans, W.P., et al.: Plasma guiding and wakefield generation for second-generation experiments. IEEE Trans. Plasma Sci. **24**(2), 331 (1996)
27. Leemans, W.P., Esarey, E.: Laser-driven plasma-wave electron accelerators. Phys. Today **62**, 44–49 (2009)
28. Schroeder, C.B., et al.: Physics considerations for laser-plasma linear colliders. Phys. Rev. ST Accel. Beams **13**, 101301 (2010)
29. Rittershofer, W., et al.: Tapered plasma channels to phase-lock accelerating and focusing forces in laser-plasma accelerators. Phys. Plasmas **17**, 063104 (2010)
30. Leemans, W.P., et al. BELLA laser and operations. In: Proceedings of NAPAC13
31. Jullien, A., et al.: Nonlinear spectral cleaning of few-cycle pulses via cross-polarized wave (XPW) generation. Appl. Phys. B **96**, 293 (2009)
32. Saleh, B.E.A., Teich, M.C.: Fundamentals of Photonics, 2nd edn. Wiley, New York (2007)
33. Born, M., Wolf, E.: Principles of Optics, 7th edn. Cambridge University Press, Cambridge (1999)
34. Sprangle, P., Krall, J., Esarey, E.: Hose-modulation instability of laser pulses in plasmas. Phys. Rev. Lett. **73**(26), 3544–3547 (1994)
35. Esarey, E., Leemans, W.P.: Nonparaxial propagation of ultrashort laser pulses in plasma channels. Phys. Rev. E **59**(1), 1082 (1999)
36. Benedetti, C., et al.: Efficient modeling of laser-plasma accelerators with INF&RNO. AIP Conf. Proc. **1299**, 250 (2010)
37. Benedetti, C., et al.: Quasi-matched propagation of ultra-short, intense laser pulses in plasma channels. Phys. Plasmas **19**, 053101 (2012)
38. van Tilborg, J.: Coherent terahertz radiation from laser-wakefield-accelerated electron beams. Ph.D. thesis, Technische Universiteit Eindhoven, Eindhoven (2006).
39. Kostyukov, I., Pukhov, A., Kiselev, S.: Phenomenological theory of laser-plasma interaction in "bubble" regime. Phys. Plasmas **11**(11), 5256 (2004)
40. Lu, W., et al.: Nonlinear theory for relativistic plasma wakefields in the blowout regime. Phys. Rev. Lett. **96**, 165002 (2006)
41. Cormier-Michel, E., et al.: Control of focusing fields in laser-plasma accelerators using higher-order modes. Phys. Rev. ST Accel. Beams **14**(3), 031303 (2011)
42. Katsouleas, T.: Physical mechanisms in the plasma wake-field accelerator. Phys. Rev. A **33**(3), 2056 (1986)
43. Sprangle, P., et al.: Wakefield generation and GeV acceleration in tapered plasma channels. Phys. Rev. E **63**(5), 056405 (2001)
44. Ting, A., et al.: Generation and measurements of high energy injection electrons from the high density laser ionization and ponderomotive acceleration. Phys. Plasmas **12**(1), 010701 (2005)
45. Pak, A., et al.: Injection and trapping of tunnel-ionized electrons into laser-produced wakes. Phys. Rev. Lett. **104**, 025003 (2010)
46. Bulanov, S.V., et al.: Transverse-wake wave breaking. Phys. Rev. Lett. **78**(22), 4205 (1997)
47. Bobrova, N.A., et al.: Simulations of a hydrogen-filled capillary discharge waveguide. Phys. Rev. E **65**, 016407 (2001)
48. Schroeder, C.B., et al.: Trapping, dark current, and wave breaking in nonlinear plasma waves. Phys. Plasmas **13**, 033103 (2006)

49. Chen, M., et al.: Electron injection and trapping in a laser wakefield by field ionization to high-charge states of gases. J. Appl. Phys. **99**, 056109 (2006)
50. Augst, S., et al.: Laser ionization of noble gases by Coulomb-barrier suppression. J. Opt. Soc. Am. B **8**(4), 858 (1991)
51. Leemans, W.P., et al.: Experiments and simulations of tunnel-ionized plasmas. Phys. Rev. A **46**(2), 1091 (1992)
52. Geddes, C. G. R.: Plasma Channel Guided Laser Wakefield Accelerator. Ph.D. thesis, University of California, Berkeley, Berkeley, CA (2005).
53. Lide, D.R. (ed.): CRC Handbook of Chemistry and Physics., 81st edn. CRC Press, Boca Raton (2000)
54. Keldysh, L.V.: Ionization in the field of a strong electromagnetic wave. Sov. Phys. JETP **20**(5), 1307 (1965)
55. Esarey, E., et al.: Self-focusing and guiding of short laser pulses in ionizaing gases and plasmas. IEEE J. Quantum Electron. **33**(11), 1879 (1997)
56. Shadwick, B.A., Schroeder, C.B., Esarey, E.: Nonlinear laser energy depletion in laser-plasma accelerators. Phys. Plasmas **16**, 056704 (2009)
57. Panasenko, D., et al.: Demonstration of a plasma mirror based on a laminar flow water film. J. Appl. Phys. **108**, 44913 (2010)
58. Sokollik, T., et al.: Tape-drive based plasma mirror. AIP Conf. Proc. **1299**, 233 (2010)
59. Broks, B.H.P., van Dijk, W., van der Mullen, J.J.A.M.: Parameter study of a pulsed capillary discharge waveguide. J. Phys. D: Appl. Phys. **39**, 2377 (2006)
60. Gonsalves, A.J., et al.: Transverse Interferometry of a hydrogen-filled capillary discharge waveguide. Phys. Rev. Lett. **98**, 025002 (2007)
61. Swamp Optics, LLC, 6300 Powers Ferry Road Suite 600–345 Atlanta, GA 30339–2919
62. Nakamura, K.: Electron Spectrometer Using the EUTERPE Magnet for TREX Staging Beam line. Lawrence Berkeley National Laboratory (2013)
63. Nakamura, K., et al.: Broadband single-shot electron spectrometer for GeV-class laser-plasma-based accelerators. Rev. Sci. Instrum. **79**, 053301 (2008)
64. Lin, C., et al.: Long-range persistence of femtosecond modulations on laser-plasma-accelerated electron beams. Phys. Rev. Lett. **108**, 094801 (2012). (Copyright 2012 by the American Physical Society)
65. Mangles, S.P.D., et al.: Laser-wakefield acceleration of monoenergetic electron beams in the first plasma-wave period. Phys. Rev. Lett. **96**(4), 215001 (2006)
66. Loos, H., et al.: Observation of coherent optical transition radiation in the LCLS linac. In: Proceedings of FEL 2008, pp. 485. Gyeongju, Korea (2008)
67. Nakamura, K., et al.: Electron beam charge diagnostics for laser plasma accelerators. Phys. Rev. ST Accel. Beams **14**(6), 062801 (2011)
68. Plateau, G.R., et al.: Low-emittance electron bunches from a laser-plasma accelerator measured using single-shot X-Ray spectroscopy. Phys. Rev. Lett. **109**, 064802 (2012). (Copyright 2012 by the American Physical Society)
69. Minty, M., Zimmermann, F.: Measurement And Control Of Charged Particle Beams. Springer, New York (2003)
70. Holloway, M.A., et al.: Multicomponent measurements of the Jefferson Lab energy recovery linac electron beam using optical transition and diffraction radiation. Phys. Rev. ST Accel. Beams **11**, 082801 (2008)
71. Yamazaki, Y., et al.: High-precision pepper-pot technique for a low-emittance electron beam. Nucl. Instrum. Methods Phys. Res. Sect. A **322**, 139 (1992)
72. Brunetti, E., et al.: Low emittance, high brilliance relativistic electron beams from a laser-plasma accelerator. Phys. Rev. Lett. **105**(19), 215007 (2010)
73. Sears, M.S.C., et al.: Emittance and divergence of laser wakefield accelerated electrons. Phys. Rev. ST Accel. Beams **13**, 092803 (2010)
74. Kneip, S., et al.: Characterization of transverse beam emittance of electrons from a laser-plasma wakefield accelerator in the bubble regime using betatron x-ray radiation. Phys. Rev. ST Accel. Beams **15**(2), 021302 (2012)

75. Esarey, E., et al.: Synchrotron radiation from electron beams in plasma-focusing channels. Phys. Rev. E **65**, 056505 (2002)
76. Benedetti, C.: Staging experiment. Presentation in the LOASIS group meeting (2010)
77. Migliorati, M., et al.: Intrinsic normalized emittance growth in laser-driven electron accelerators. Phys. Rev. ST Accel. Beams **16**(1), 011302 (2013)
78. Vay, J.-L., et al.: Computer simulation of multiple consecutive laser-plasma acceleration stages. 54th Annual meeting of the APS Division of Plasma Physics (2012)
79. Vay, J.-L., et al.: Simulations of laser plasma accelerators. DOE HEP Review (2013)
80. Stuart, B.C., et al.: Laser-induced damage in dielectrics with nanosecond to subpicosecond pulses. Phys. Rev. Lett. **74**(12), 2248 (1995)
81. Doumy, G., et al.: Complete characterization of a plasma mirror for the production of high-contrast ultraintense laser pulses. Phys. Rev. E **69**, 026402 (2004)
82. Gold, D.M.: Direct measurement of prepulse suppression by use of a plasma shutter. Opt. Lett. **19**(23), 2006 (1994)
83. Dromey, B., et al.: The plasma mirror–a subpicosecond optical switch for ultrahigh power lasers. Rev. Sci. Instrum. **75**, 3 (2004)
84. The Physics of Laser Plasma Interactions. Westview Press, Boulder (2003)
85. Milchberg, H.M., Freeman, R.R.: Light absorption in ultrashort scale length plasmas. J. Opt. Soc. Am. B **6**(7), 1351 (1989)
86. Eliezer, S.: The Interaction of High-Power Lasers with Plasmas. Plasma Physics. IOP Publishing Ltd, Philadelphia (2002)
87. Thaury, C., et al.: Plasma mirrors for ultrahigh-intensity optics. Nat. Phys. **3**, 424 (2007)
88. Thaury, C., Quere, F.: High-order harmonic and attosecond pulse generation on plasma mirrors: basic mechanisms. J. Phys. B: At. Mol. Opt. Phys. **43**, 213001 (2010)
89. Micromap Corporation, Tucson, AZ; phone +1 520–882-1911
90. Bhushan, B.: Introduction to Tribology, 2nd edn. Wiley Online, New York (2013)
91. Shaw, B.H., et al.: High-peak-power surface high-harmonic generation at extreme ultraviolet wavelengths from a tape. J. Appl. Phys. **114**(4), 043106 (2013)
92. Reid, M.B.: Electron beam emittance growth in thin foils: a betatron functionanalysis. J. Appl. Phys. **70**(11), 7185 (1991)
93. Jackson, E.A.: Nonlinear oscillations in a cold plasma. Phys. Fluids **3**(5), 831 (1960)
94. Siders, W., et al.: Laser wakefield excitation and measurement by femtosecond longitudinal interferometry. Phys. Rev. Lett. **76**(19), 3570 (1996)
95. Marques, J.R., et al.: Laser wakefield: experimental study of nonlinear radial electron oscillations. Phys. Plasmas **5**(4), 1162 (1998)
96. Matlis, N.H., et al.: Snapshots of laser wakefields. Nat. Phys. **2**(11), 749 (2006)
97. Radiant Zemax, 22908 NE Alder Crest Drive, Suite 100 Redmond, WA 98053 USA
98. PHASICS S. A., XTEC Bât. 404, Campus de l'Ecole Polytechnique, Route de Saclay, 91128 Palaiseau, FRANCE
99. Pampaloni, F., Enderlein, J.: Gaussian, Hermite-Gaussian, and Laguerre-Gaussian beams: A primer (2004). arXiv:physics/0410021
100. Reprinted with permission from Shiraishi, S., et al.: Laser red shifting based characterization of wakefield excitation in a laser-plasma accelerator. Phys. Plasmas **20**, 6 (2013) 063103. (Copyright 2013, American Institue of Physics)
101. Schroeder, C.B., et al.: Group velocity and pulse lengthening of mismatched laser pulses in plasma channels. Phys. Plasmas **18**, 083103 (2011)
102. Leemans, W.P., et al.: Electron-yield enhancement in a laser-wakefield accelerator driven by asymmetric laser pulses. Phys. Rev. Lett. **89**(17), 174802 (2002)
103. Schroeder, C.B., et al.: Frequency chirp and pulse shape effects in self-modulated laser wakefield accelerators. Phys. Plasmas **10**(5), 2039 (2003)
104. Tóth, C., et al.: Tuning of laser pulse shapes in grating-based compressors for optimal electron acceleration in plasmas. Opt. Lett. **28**(19), 1823 (2003)
111. Reprinted with permission from Gonsalves, A.J., et al.: Plasma channel diagnostic based on laser centroid oscillations. Phys. Plasmas **17**, 5. (Copyright 2010, American Institute of Physics)

106. Spence, D.J., Hooker, S.M.: Investigation of a hydrogen plasma waveguide. Phys. Rev. E **63**, 015401 (2000)
107. Gonsalves, A. J.: Investigation of a hydrogen-filled capillary discharge waveguide for laser-driven plasma accelerators. Ph.D. thesis, University of Oxford (2006)
108. Broks, B.H.P., Garloff, K., van der Mullen, J.J.A.M.: Nonlocal-thermal-equilibrium model of a pulsed capillary discharge waveguide. Phys. Rev. E **71**(1), 016401 (2005)
109. Murphy, C.D., et al.: Evidence of photon acceleration by laser wake fields. Phys. Plasmas **13**, 033108 (2006)
110. Andreev, N.E., et al.: Analysis of laser wakefield dynamics in capillary tubes. New J. Phys. **12**, 045024 (2010)
111. Schroeder, C.B., et al.: Nonlinear pulse propagation and phase velocity of laser-driven plasma waves. Phys. Rev. Lett. **106**, 135002 (2011)
112. Hamamatsu Photonics, 360 Foothill Road, Box 6910 Bridgewater, NJ 08807
113. ANSYS Inc, Southpointe 275 Technology Drive Canonsburg, PA 15317
114. Chen, F.F.: Introduction to Plasma Physics and Controlled Fusion, 2nd edn. Plenum Press, New York (1984)
115. Byrd, J.M., et al. : Femtosecond synchronization of laser systems for the LCLS. Proceedings of IPAC'10 (2010)
116. Huang, G., et al.: LCLS Femto-second timing and synchornization system update. Proceedings of IPAC2012, p. 2176
117. Adair, R., Chase, L.L., Payne, S.A.: Nonlinear refractive index of optical crystals. Phys. Rev. B **39**(5), 3337 (1989)
118. Fill, E.E.: Focusing limits of ultrashort laser pulses: analytical theory. J. Opt. Soc. Am. B **11**(11), 2241 (1994)

Zeitfracht Medien GmbH
Ferdinand-Jühlke-Straße 7
99095 Erfurt, Deutschland
produktsicherheit@kolibri360.de